Relaxed Abduction

Thomas Hubauer

Relaxed Abduction

Robust Information Interpretation
for Industrial Applications

 Springer Vieweg

Thomas Hubauer
München, Deutschland

Dissertation at the University of Lübeck, 2015

ISBN 978-3-658-14406-7   ISBN 978-3-658-14407-4 (eBook)
DOI 10.1007/978-3-658-14407-4

Library of Congress Control Number: 2016942010

Springer Vieweg

Printed on acid-free paper

This Springer Vieweg imprint is published by Springer Nature
The registered company is Springer Fachmedien Wiesbaden GmbH

I owe thanks to a number of people whose continued support made this thesis possible. First and foremost, I would like to express my gratefulness to my family for their encouragement and backing during these years. My thanks extend to Prof. Dr. Ralf Möller for supervising and teaching me as an external PhD candidate over many years, and to Prof. Dr. Till Tantau for his readiness to review my thesis. Last in order but definitely not least in importance, my gratitude goes to Dr. Özgür Özcep, Dr. Steffen Lamparter, and Dr. Stephan Grimm for their criticism, suggestions, and continued encouragement.

Preface

Automated information interpretation is gaining momentum in industrial applications. One of the major challenges in this context is the appropriate treatment of incompleteness of both the observations about the world, and the domain model formalizing it. This dissertation proposes a novel approach called "Relaxed Abduction", which is able to provide reasonable interpretations even if it would not be possible or extremely complex to explain all observations made. The approach is based on the idea of treating explanatory power and consilience of an interpretation bi-criterially instead of mapping them onto a one-dimensional scale. Based on a formalization of the proposed approach, the thesis investigates concrete instantiations and their properties (particularly runtime complexity), and proposes an extension which allows to handle incoming changes in the underlying data incrementally. Two algorithms for solving relaxed abduction problems are proposed (one generic and one $\mathcal{EL}$-specific) and evaluated in a real-world use case, confirming the theoretical results in practice. Additionally, to close the gap between academic research and industrial applications, the thesis proposes a methodology to structure diagnosis problems according to ISO 13379 and express it by means of several description logic knowledge bases. The dissertation results show that the proposed, flexible notion of abduction is indeed novel, relevant for industry, and permits practical usage. Future steps required to bring relaxed abduction into application are identified.

Contents

List of Figures

List of Tables

Other Listings

List of Examples

List of Definitions

List of Theorems

List of Algorithms

1 Introduction

1.1 Motivation

The Semantic Web vision in its purest form is to make the World Wide Web machine-understandable. The Semantic Web community wants to realize this vision by annotating all types of resources, e. g. text or hypertext documents, images, and audio files with metadata that can be processed automatically by computers. This information processing is facilitated by technologies such as ontologies providing background information about an ever growing number of domains, and automated reasoning employed to evaluate information represented using such a formalisation of the domain. Although the ultimate goal of having computers automatically search the web and provide the interpreted results to the user has not been achieved yet, current developments like the introduction of virtual personal agents into smartphones show that the knowledge representation and reasoning (KR&R) technologies underpinning the Semantic Web have reached an impressive level of maturity.

Over the last years, KR&R technologies have found application in fields far beyond this original vision, such as the analysis of multimedia data, diagnostic services for complex industrial systems, or systems for supporting elderly people in their home environment, to name only a few. Despite their apparent diversity, all these applications can be understood as instances of a more general information interpretation task: Given potentially large amounts of observational data (often acquired from sensors including "intelligent sensors" that provide preprocessed data) and a description of the laws governing the specific domain, the goal is to determine a situation description which explains these observations. Declarative descriptions of the domain with formal semantics (often called semantic models or ontologies) are conveniently used to express the underlying relationships between situations and observations as well as additional background knowledge on the domain, for instance object descriptions, structural information on machinery, or the layout of an elderly person's home. In this domain of *model-based information interpretation*, the notion of a model is frequently used to designate both a

formal representation of a domain and an interpretation of information in
the sense of model theoretic semantics. In this thesis, we adopt this general
approach, however, in places where confusion between the two meanings
might arise otherwise, we employ terms such as "knowledge base" or "domain
formalization" to designate the first interpretation.

The most prominent reasoning paradigm applied in KR&R and the Semantic Web is *deduction*: based on a set of axioms, determine what additional
information is entailed. Deductive inference is well-studied and virtually all
available reasoning tools are based on deduction, mainly due to the fact that
deduction always yields correct conclusions: as long as the input is free of
contradictions, deduction will never lead to any invalid conclusions.[1] On the
downside, deduction is only applicable in situations where all relevant input
information is known, or can at least be estimated. A significant proportion
of information interpretation tasks however do not comply with this prerequisite, reasons for this being both fundamental (e. g. non-observability of
internal processes) and practical (e. g. sensor failure). *Abductive reasoning*
provides a solution to this challenge by deriving *hypothetical explanations*
for set of dedicated axioms (e. g. representing observations), and allowing
a user to test these potential solutions against new information later on.
This capability of handling incomplete observational data makes abduction
a powerful tool for most information interpretation tasks. However, weak or
faulty domain formalisations still pose a significant challenge to abductive
information interpretation, particularly since they may make it impossible
to explain individual observations. Whereas some early approaches fail to
find any solution at all once a single observation cannot be accounted for,
more recent methods have found solutions to address this common issue –
typically by allowing to simply assume observations which cannot be explained otherwise. These approaches inherently incur a tradeoff between
the "cost" of assuming missing information and the "benefit" of explaining a
certain observation. Despite their obvious appeal, all these solutions either
return overly complex, suboptimal, or incomplete solutions.[2] This obviously
poses a significant burden on the use of abductive reasoning in addressing
real-world information interpretation problems.

This thesis reports on research in bringing abductive reasoning to industrial
application. It proposes a novel, flexible method for logic-based abduction

---

[1] Obviously, whether this entailed information indeed matches the intuitive expectations
of a user hinges on the fact that the formalisation correctly captures reality. It is
therefore no surprise that, depending on the complexity of the domain to be captured
and the degree of modelling detail, such formalisations tend to be very large.

[2] We discuss this topic in more detail in Section 3.2.1.

over imperfect domain formalisations that avoids the drawbacks outlined before, and shows how it can be implemented efficiently. The applicability of the proposed method is investigated in two industrial use cases on diagnostics of machinery and rule bases respectively. The domain formalizations of both use cases can be understood to instantiate a common meta model which was developed in accordance with established industry norms on industrial diagnostics.

1.2 Research Objectives

The thesis addresses the challenge of bringing logic engineering in general and, more specifically, abductive inference into industrial information interpretation applications. Our main research objective is therefore to *develop a method for model-based information interpretation that addresses both observational and domain model incompleteness, can be practically implemented, and easily applied in a wide range of industrial use cases.*

First and foremost, we aim at providing a formally founded framework for information interpretation that guarantees both soundness and completeness of the result. This framework must therefore meet the following requirements:

R1 (Expressiveness)
To allow for a compact representation of information and verifiability of results, the framework must support modelling of and reasoning over complex relational structures based on formal semantics.

R2 (Robustness to observational incompleteness)
The framework must support devising analysis results of a situation even if the observational state is incomplete (meaning that certain observations characterising a situation are unknown, e. g. due to sensor failure).

R3 (Robustness to imperfect formalisations)
The framework must support devising analysis results of a situation with respect to incomplete or faulty domain representations (e. g. due to deficient expert knowledge on the domain under consideration).

R4 (Optimality)
To minimize distraction, the framework must not return suboptimal solutions (with respect to user-definable optimality criteria).

R5 (Coverage)

To maximize information gain, the framework must explore the full spectrum of solutions in the area of tension between explanation complexity and observation coverage.

Such a formal framework, however, defines characteristics of both the domain representation and the solutions on a rather abstract level. To be of practical use, it needs to be complemented with an algorithm for effective and efficient generation of solutions. Requirements for this algorithm are:

R6 (Correctness)

The algorithm must be sound, complete, and computable in finite time.

R7 (Efficiency)

The algorithm must allow for an efficient implementation. (Note that in practice this does not necessarily require that the algorithm has polynomial complexity. Numerous examples show that algorithms with well-behaved exponential runtime can be applied very successfully in practice.)

Taken together, these requirements characterise an information interpretation methodology that allows to make best use of (potentially incomplete) information in a correct and efficient manner. Similar requirements can be found e. g. in Dressler & Puppe (1999), stressing the relevance of the criteria listed above. The knowledge-based nature makes the proposed solution apt for a wide range of practical applications including, but not limited to, diagnostic tasks in industry. We now introduce a simple diagnostics scenario that will serve as a motivating example throughout this thesis. Next, the subsequent section gives a short summary on how this work contributes to a number of scientific fields.

Example 1.1 (A simple diagnostics scenario)

In our motivating example, we consider a simplified production system equipped with a diagnostic unit. The system under consideration comprises at least two independent subsystems, namely communications and transportation. The former provides PROFINET connectivity to other devices in the factory, the latter a mechanical gripper that can be used to move products from the storage location onto the internal conveyor belt. The whole system is governed by a central main control unit (or MCU, for short). Experts working with the system have observed that a fluctuating power supply manifests by intermittent outages of the main control unit while the communication links remain functional and the mechanical gripper of the production system is unaffected (the observations entailed by the diagnosis). Similarly, the maintenance manual of the system states that bugs in the MCU software typically lead to intermittent MCU outages as well, accompanied by irregular (i. e. unintended) derivations in the movement patterns of both gripper and conveyor belt.

To minimize downtimes, it is essential to identify faults quickly and reliably based on the available measurements from sensors attached to the system. This task is to be accomplished by the diagnostic unit, based on automated (abductive) reasoning.

1.3 Scientific Contributions

To address Requirements **R1** to **R7**, this thesis contributes to the fields of logic-based reasoning (with focus on description logics), knowledge representation, and industrial application of logic engineering. More specifically, our contributions include the following:

- We identify the adequate *treatment of imperfect domain formalisations* as an inherent flaw of standard formulations of logic-based abduction.
- We introduce *relaxed abduction* as a natural extension of abduction that is capable of simultaneously addressing incompleteness of observations and domain representation in a sound and significantly more flexible way than existing solutions do.
- We analyse the *formal properties* of this reasoning method, clarify its relation to standard abduction, and investigate the effects of different criteria for solution optimality.
- We show how relaxed abduction problems can be solved both for general description logics and – using a more involved algorithm – for the tractable description logic $\mathcal{EL}^+$, and analyse the respective computational complexity of the algorithms provided.
- We examine *extensions* to the $\mathcal{EL}^+$ algorithm, namely possible extension to more expressive modelling languages as well as efficient support for the addition and retraction of axioms.
- We investigate how different types of information interpretation problems can be *mapped formally* into the relaxed abduction framework, thereby strengthening the link between our novel formalism and real industrial applications.

Summarising our contributions, this thesis makes a significant step in closing the gap between theoretical results in the field of knowledge representation and reasoning, and their practical applicability in the field of industrial information interpretation. By formally relaxing the well-studied formalism of logic-based abduction, we significantly extend its applicability in real-world applications without sacrificing its solid theoretical foundations. Moreover, our studies on extensions and on a formal mapping to an ISO-standardised diagnostic vocabulary help tighten the bound between theory and successful application. Activities concerned with the dissemination of the results developed in the course of this thesis are presented in the subsequent section.

1.4 Dissemination Activities

This section summarises activities connected to the dissemination of the results provided by this thesis. The presentation is clustered by types of contributions.

Conference Papers

- Hubauer et al. (2011b) introduces an agent-based system for flexible manufacturing control using a diagnostics unit based on relaxed abduction. The results contributed mostly to the use case presentations in Sections 4.3.1 and 4.3.2.
- Legat et al. (2011) complements Hubauer et al. (2011b) with in-depth information on the formalisms used for automatic assessment of remaining machine capabilities.
- Hubauer et al. (2012) presents a conceptual model for diagnostics motivated by ISO standards, and a mapping between this formalization and the proposed notion of relaxed abduction. It furthermore provides evaluation results in the context of the turbine diagnostics use case. The contents of this paper contributed to Sections 3.1, 4.2 and 4.3.1.
- Grimm et al. (2012) proposes an approach for bringing completion-based reasoning to embedded devices such as programmable logic controllers (PLCs), where computations must adhere to strict cycle time restrictions. Though currently limited to classification of $\mathcal{EL}^{++}$ ontologies, the results gave valuable input to the implementation of the RAbIT system presented in Section 4.1 and may serve as a starting point for future work in the direction of embedded (relaxed) abduction.

Workshop Papers

- Hubauer et al. (2010) focuses on standard logic-based abduction in the context of description logics. It laid foundations for the hypergraph-motivated analysis of abduction and, based on that, relaxed abduction.
- Hubauer et al. (2011a) motivates and introduces the framework of relaxed abduction. It provides a first analysis of its properties as well as the algorithm for solving relaxed abduction for $\mathcal{EL}^+$ and first complexity results (which have been tightened meanwhile). This publication mainly contributed to Sections 3.2 and 3.3.2.

Other Results

- The algorithm for solving relaxed abduction over $\mathcal{EL}^{++}$ knowledge bases has been implemented in a Java component named RAbIT (an acronym for *R*obust *A*bductive *I*nference *T*ool).

1.5 Outline

The remainder of this thesis is structured as follows: In Chapter 2 we introduce basics on ontologies and description logics, abductive reasoning, order theory, and hypergraphs. Chapter 3 steps right into the core contributions of this thesis: After introducing a DL-based formalisation of diagnostics (Section 3.1), we present the framework of relaxed abduction in general (Section 3.2) as well as our algorithms for solving relaxed abduction problems in the context of description logics (Section 3.3). Moreover, extensions to the basic setting are considered in Section 3.4, and the chapter concludes with a comparison to related approaches in Section 3.5. Chapter 4 presents the RAbIT system, our implementation of the $\mathcal{EL}^+$-based algorithm for solving relaxed abduction, and analyses its performance (Sections 4.1 and 4.2). Next, two case studies on applying relaxed abduction to problems from the diagnostics domain are presented in Section 4.3. We conclude in Chapter 5 with a summary of our results and interesting open research questions.

2 Preliminaries

This chapter introduces topics that are relevant in the course of this thesis; the presentations are hereby restricted to the aspects required in the remainder of this work and not intended to be complete. Where needed to simplify presentation later on, we also introduce new terminology and prove related lemmas to simplify proofs later on. Section 2.1 gives a short overview on ontologies and their relation to description logics as well as classical reasoning tasks for DLs. Section 2.2 complements this account with an introduction to abductive reasoning both from a general perspective and in the context of description logics. Alternative definitions of abduction from literature are introduced, and their relation to the definition given here is formally analysed. As solving certain classes of relaxed abduction problems will be reduced to finding optimal hyperpaths later on, a number of relevant order-theoretic notions is introduced in Section 2.3; a basic account on hypergraphs, an extension of standard graphs, can be found in Section 2.4.

2.1 Ontologies and Description Logics

Knowledge about the entities of a domain and their numerous interrelations can conveniently be captured in an explicit and intelligible way by means of an *ontology*. Representing domains by means of ontologies provides a deeper understanding of the domain as compared to subsymbolic approaches or other representations acquired by learning that often do not support the representation of relational information. This makes it possible for domain experts to gradually extend and adapt the formalisation to integrate their knowledge. Nowadays, the method of choice for representing ontologies is the *Web Ontology Language (OWL)* which is semantically founded on the description logic (DL) formalism described in Nardi et al. (2003).

Description logics are a family of formal knowledge representation formalisms and can be understood as fragments of first-order logic restricted to at most binary predicates. The main representational elements are *concepts* (unary predicates), *roles* (binary predicates), and *individuals* (terms). Description logics have model-theoretic semantics defined by means of an interpretation function $\cdot^{\mathcal{I}}$ that maps individuals to elements of the domain Δ, concepts to sets of elements, and roles to sets of element pairs. The entailment relation between a set X of DL axioms and a single DL axiom x, denoted $X \models x$, is defined as usual. For two sets of axioms X, Y, we let $X \models Y$ stand for $X \models \bigwedge Y$ or equivalently: $X \models y$ for all $y \in Y$. $X \models\!|Y$ stands for $Y \models X$, and $X \equiv Y$ is an abbreviation for $X \models\!|Y$ and $X \models Y$. Later in this thesis, we need the intuition of an axiom set not containing any redundant axioms. To capture this concept, we define the notion of a *logically independent set of axioms* as follows:

Definition 2.1 (Logically independent axiom set)
Let X be a set of axioms with subsets S and S'. Then X is logically independent *if and only if $S' \models S$ implies $S \subseteq S'$.*

In other words, an axiom set X is logically independent if *no* subset $S \subseteq X$ is redundant in the sense specified by Grimm & Wissmann (2011).

A concrete description logic is characterized by the constructors available for forming complex concepts from simple ones, and types of *axioms* that can be stated. This choice typically constitutes a tradeoff between expressiveness of the resulting language and computational complexity of deciding satisfiability of a set of axioms called a *knowledge base* (KB) or *ontology*. This task also known as *consistency checking* is one of the central reasoning tasks in description logics, its decidability is a property shared by all DLs. There is a number of other *standard inference tasks* which can be solved by reduction to consistency checking for most description logics[1]: *subsumption checking* (determine whether every instance of concept C must necessarily be an instance of concept D as well, denoted $C \sqsubseteq D$), *classification* (determine all subsumption relations between concepts of the knowledge base)[2], *instance checking* (determine whether individual i is an instance of a concept C, denoted $C(i)$ or $i : C$), and *realization* (for every individual of the knowledge base determine the most specific concept names it is an instance of).

---

[1] Basically, reducibility for concept-oriented reasoning tasks rests on the capability of expressing disjointness of concepts, whereas reducibility of individual-oriented tasks also requires the existence of nominals, or enumerated concepts.

[2] It should be pointed out that this definition coming from the community around lightweight description logics differs from the standard definition used in most other parts of the DL community, where classification determines only the most specific subsumption relations between any pair of concepts. The difference can be understood as an additional filter step on the full set of all subsumptions, retaining only the most specific ones. Therefore, classification according to the standard definition is typically somewhat more expensive from a computational perspective.

In its second version OWL 2 (W3C OWL Working Group, 2009a), the Web Ontology Language is largely based on the highly expressive description logic $\mathcal{SROIQ}$ (Horrocks et al., 2006). $\mathcal{SROIQ}$ provides many powerful features including enumerated classes $\{i_1, \ldots, i_k\}$, full negation $\neg C$, various qualified restrictions, role hierarchies, and inverse roles. This however leads to high computational cost of the reasoning tasks introduced before. Deciding concept satisfiability was shown to be N2ExpTime-complete in Kazakov (2008). Since the doubly exponential complexity of $\mathcal{SROIQ}$ may be prohibitive for certain applications, OWL 2 comes with a number of so-called profiles which offer language variants with reduced expressiveness and better computational properties. One of these profiles is OWL 2 EL, which has been shown to be especially well suited for the representation of large taxonomic structures such as the well-known SNOMED-CT[3] and GALEN[4] medical ontologies. The EL profile is based on the description logic $\mathcal{EL}^{++}$ (Baader et al., 2005a), for which consistency is decidable in PTime. This significant reduction of complexity is made possible by the absence of a number of constructors such as universal quantification, full negation and disjointness axioms for concepts and roles, along with a number of other constructs that could be used to express one of the previous. It has been shown by Baader et al. that $\mathcal{EL}^{++}$ is maximal in the sense that none of these dropped features can be included without sacrificing tractability. Despite these syntactic restrictions, $\mathcal{EL}^{++}$ supports expressing information about individuals by means of nominals (concepts comprised of exactly one named individual), which allows to reduce reasoning over individuals to standard concept-level reasoning. The same complexity bounds hold for the sublanguage $\mathcal{EL}^+$ which has been proved sufficient for the representation of diagnostic ontologies in various contexts (see e. g. Baader et al. (2007)). An $\mathcal{EL}^+$ ontology $\mathcal{T}$ is a set of concept inclusion axioms of the form Turbine $\sqcap \exists$ propelledBy . Gas $\sqsubseteq$ GasTurbine (which states that a gas turbine is a turbine propelled by gas) and role inclusion axioms of the form directlyControls $\circ$ hasSubComponent $\sqsubseteq$ controls (stating that control is propagated over the partonomy of a system). Table 2.1 summarizes the syntax and semantics of the concept constructors and the axioms that can be represented in $\mathcal{EL}^+$ (where A is an atomic concept, C, D are arbitrary concepts, and r_i, r, s are roles). For details on the description logic constructors we refer the reader to Nardi et al. (2003); W3C OWL Working Group (2009b).

For making explicit that an axiom ax is expressed in the description logic

---

[3]http://www.nlm.nih.gov/research/umls/Snomed/snomed_main.html
[4]http://www.openclinical.org/prj_galen.html

Table 2.1: $\mathcal{EL}^+$ syntax and semantics

	Syntax	Semantics
top concept	$\top$	$\Delta^{\mathcal{I}}$
atomic concept	A	$A^{\mathcal{I}}$
concept conjunction	$C \sqcap D$	$C^{\mathcal{I}} \cap D^{\mathcal{I}}$
existential restriction	$\exists r . C$	$\{x \in \Delta^{\mathcal{I}} \mid \exists y \in \Delta^{\mathcal{I}} :$ $(x, y) \in r^{\mathcal{I}} \wedge y \in C^{\mathcal{I}}\}$
concept inclusion axiom	$C \sqsubseteq D$	$C^{\mathcal{I}} \subseteq D^{\mathcal{I}}$
role inclusion axiom	$r_1 \circ \cdots \circ r_n \sqsubseteq s$	$r_1^{\mathcal{I}} \circ \cdots \circ r_n^{\mathcal{I}} \subseteq s^{\mathcal{I}}$

L we sometimes call ax an *L-axiom*. Analogously, a *L-knowledge base* is a set of L-axioms. Given a description logic knowledge base $\mathcal{T}$, we denote the set of all role names occurring in $\mathcal{T}$ by N_{R}, the set of all concept names by N_{C}, and define $N_{\mathrm{C}}^{\top} = N_{\mathrm{C}} \cup \{\top\}$.

2.2 Abductive Reasoning

This section provides an overview on abductive reasoning. We first give a general introduction on the framework of (axiom-based) abduction and its instantiations. Next, we relate this definition to a relatively new formulation called concept abduction and show how the former subsumes the latter. The section concludes with an analysis of the shortcomings of logic-based abduction with respect to the processing of deficient domain representations, one of the main challenges addressed in this thesis.

2.2.1 Standard Notions of Abduction

Abduction was introduced in philosophy of logic in the late 19th century by Charles Sanders Pierce, who characterized it as the sole reasoning method generating new information (Hartshorne & Weiss, 1931). It was rediscovered as a method for determining diagnoses in the 1970s by Pople (Pople, 1973), and has since then been employed as a method for interpreting incomplete information in various applications such as text interpretation (Hobbs et al., 1993), plan generation and analysis (Appelt & Pollack, 1992), and analysis of sensor (Shanahan, 2005) or multimedia data (Möller & Neumann, 2008; Peraldi et al., 2007). In the spirit of Pierce, abduction can be seen as an inference scheme capable of providing possible explanations for some observation. It is conveniently represented by the rule

$$\frac{\phi \supset \omega \qquad \omega'}{\phi\Theta}$$

and interpreted as an inversion of the well-known generalised modus ponens rule. This new derivation rule allows to derive $\phi\Theta$ as a hypothetical explanation for the occurrence of ω', given that the presence of ϕ in some sense justifies ω (where Θ is a unifier[5] for ω and ω'). Such explanations need not be unique, and while each single explanation is typically required to fulfil some consistency requirement, different competing explanations may well be contradictory to each other. Hypotheses can be falsified by adding information (e. g. representing additional observations), making abduction an inherently non-monotonic reasoning scheme.

Note that this general notion of abduction does not presuppose any causality between ϕ and ω, as indicated by the notation $\phi \supset \omega$ of the *major premise*. On the contrary, various notions of how ϕ sanctions the presence of ω give rise to different definitions of abduction as explicated in detail in Paul (1993):

- *Set-cover abduction* implements the major hypothesis $\phi \supset \omega$ by means of an explicit mapping $e : \Phi \mapsto \Omega$ called *explanatory power* that relates sets of hypotheses to the sets of observations they explain. Determining abductive solutions can then be reduced to finding minimum covers of the observation set, based on the explanatory power. For realistic examples however, the mapping e is prone to get huge and it is unlikely that it can be defined correctly by hand, let alone updated on changes in the domain.

- *Logic-based abduction* can be understood to replace the explicit mapping used in the set-cover approach by a knowledge base $\mathcal{T}$, making the full power of the formal language available for representing expressive yet concise mappings. In addition to the domain formalisation, a set $\mathcal{A}$ of abducible axioms is provided. A logic-based abduction problem is solved by determining subsets of $\mathcal{A}$ which are consistent with the axiom set $\mathcal{T}$ and explain the observation(s) given $\mathcal{T}$. Hypothesis construction for logic-based abduction is typically more involved than in the set-cover variant since the mapping e is only known implicitly.

---

[5]Given terms $\omega_1, \ldots, \omega_n$, a *substitution* is a mapping from variables to terms which can be understood as a generalised type of renaming, replacing (some of the) variables with potentially complex terms comprising constants, variables, and functions. A unifier Θ for $\omega_1, \ldots, \omega_n$ is a special substitution such that $\omega_1\Theta = \ldots = \omega_n\Theta$.

- *Knowledge-level abduction* generalizes logic-based abduction with a conceptual model of belief. In a nutshell, ϕ explains ω w.r.t. an epistemic state if (in this state) it is believed that $\phi \to \omega$, and the negation of ϕ is not believed. This approach is claimed to be relevant e.g. for situated agents where the epistemic state might change regularly due to external factors.

Obviously, set-cover abduction will typically be too weak to meet Requirement **R1**. As representation of and reasoning over belief is not relevant for the application scenarios at hand, we focus on logic-based reasoning over L-knowledge bases (for decidable logics L), i.e. we understand abduction as a *non-standard inference task* over formal, logic-based domain representations. In this context, an abduction problem can be defined as follows (c.f. Eiter & Gottlob, 1995):

Definition 2.2 (Abduction problem)
An abduction problem *is a 3-tuple* $\mathbf{AP} = (\mathcal{T}, \mathcal{A}, \mathcal{O})$, *where*

- $\mathcal{T}$ *is a set of axioms that formalise the domain,*
- $\mathcal{A}$ *is a set of abducible axioms representing assumptions, and*
- $\mathcal{O}$ *is a set of axioms representing observations.*

A solution *to* $\mathbf{AP}$ *is a set* $A \subseteq \mathcal{A}$ *such that* $\mathcal{T} \cup A$ *is consistent and* $\mathcal{T} \cup A \models \mathcal{O}$. *Naturally extending the standard notation, we write* $A \models \mathbf{AP}$ *to state that* A *solves* $\mathbf{AP}$. *The* solution set *for* $\mathbf{AP}$ *is defined by* $Sol_{\mathbf{AP}} := \{A \mid A \models \mathbf{AP}\}$.

It should be pointed out that abduction does not at all require that every conclusion of $\mathcal{T} \cup A$ has been observed (i.e. is an element of $\mathcal{O}$). This separates abduction from mere backward-chaining rule-based deduction, and enables it to flexibly address situations where the observation set is incomplete.[6]

Due to the fact that an abduction problem does in general not have one unique solution A but a collection of alternative answers $A_1, \ldots, A_k$, one typically selects optimal solutions by means of a (not necessarily total) preference order $\preceq$, with $A_i \preceq A_j$ expressing that A_i is at least as good as A_j (see Section 2.3 for an introduction to orders). Abductive inference can then conveniently be formulated as an optimization problem as shown next:

---

[6]Backward-chaining rule-based reasoning can obviously emulate this feature by breaking down rules with multiple conjuncts in the precedent. However, this typically leads to a significant explosion of the number of rules to consider.

Definition 2.3 (Preferential abduction problem)
A preferential abduction problem is a 4-tuple $\mathbf{PrefAP} = (\mathcal{T}, \mathcal{A}, \mathcal{O}, \preceq_\mathcal{A})$, *where*

- $\mathcal{T}$ *is a set of axioms that formalise the domain,*
- $\mathcal{A}$ *is a set of abducible axioms representing assumptions,*
- $\mathcal{O}$ *is a set of axioms representing observations, and*
- $\preceq_\mathcal{A} \subseteq \mathcal{P}(\mathcal{A}) \times \mathcal{P}(\mathcal{A})$ *is a partial order over sets of assumptions.*

A pre-solution to $\mathbf{PrefAP}$ *is a set* $A \subseteq \mathcal{A}$ *such that* $\mathcal{T} \cup A$ *is consistent and* $\mathcal{T} \cup A \models \mathcal{O}$. *A solution is an* $\preceq_\mathcal{A}$*-minimal element of the set of pre-solutions.* *We write* $A \models \mathbf{PrefAP}$ *to state that* A *solves* $\mathbf{PrefAP}$. *The solution set for* $\mathbf{PrefAP}$ *is defined by* $Sol_{\mathbf{PrefAP}} := \{A \mid A \models \mathbf{PrefAP}\}$.

Typical orders over sets include subset minimality ($A_i \preceq^s A_j$ if and only if $A_i \subseteq A_j$), minimum cardinality ($A_i \preceq^c A_j$ if and only if $|A_i| \leq |A_j|$), and weight-based approaches defined by a function w that assigns numerical weights to subsets of $\mathcal{A}$ (i.e. $A_i \preceq^w A_j$ if and only if $w(A_i) \leq w(A_j)$).

2.2.2 Concept-based Notions of Abduction

A different formulation of abduction due to Bienvenu (2008) employs atomic concepts for representing the abducibles and the observation. Note that we use the term *subsumption-based abduction* for this approach instead of the original name *concept-based abduction*, in order to clearly distinguish it from *concept abduction* introduced later. For similar reasons, we will from now on use the term axiom-based abduction for the standard definition (Definition 2.2). After introducing subsumption-based abduction in Definition 2.4, we show in Lemma 2.1 that this notion is less general than axiom abduction.

Definition 2.4 (Subsumption-based abduction problem)
A subsumption-based abduction problem is a 3-tuple $\mathbf{SAP} = (\mathcal{T}, \mathcal{H}, \mathsf{O})$, *where*

- $\mathcal{T}$ *is a set of axioms that formalise the domain,*
- $\mathcal{H}$ *is a set of atomic concepts (each representing a possible assumption), and*
- O *is an atomic concept representing the observation.*

A solution to $\mathbf{SAP}$ *is a set* $H \subseteq \mathcal{H}$ *such that* $\sqcap_{\mathsf{C}_i \in H} \mathsf{C}_i$ *is consistent w. r. t.* $\mathcal{T}$ *and* $\mathcal{T} \models \sqcap_{\mathsf{C}_i \in H} \mathsf{C}_i \sqsubseteq \mathsf{O}$.

Lemma 2.1 (Axiom-based can simulate subsumption-based abd.)
A subsumption-based abduction problem **SAP** $- (\mathcal{T}, \mathcal{H}, \mathsf{O})$ *can be solved by determining solutions to the corresponding axiom-based abduction problem* **AP** $= (\mathcal{T}, \mathcal{A}, \mathcal{O})$ *defined by:*

- $\mathcal{A} = \{\mathsf{C}^* \sqsubseteq \mathsf{C}_i \mid \mathsf{C}_i \in \mathcal{H}\},$
- $\mathcal{O} = \{\mathsf{C}^* \sqsubseteq \mathsf{O}\},$ *and*
- C^* *is a new concept name,* $\mathsf{C}^* \notin N_{\mathrm{C}}$

More concretely, $A = \{\mathsf{C}^* \sqsubseteq \mathsf{C}_i\}$ *solves* **AP** *if* $H = \{\mathsf{C}_i \mid \mathsf{C}^* \sqsubseteq \mathsf{C}_i \in A\}$ *solves* **SAP**.

Proof. Let H solve **SAP**, i.e. $\mathcal{T} \models \bigsqcap_{\mathsf{C}_i \in H} \mathsf{C}_i \sqsubseteq \mathsf{O}$. Then $\mathcal{T} \equiv \mathcal{T} \cup \{\bigsqcap_{\mathsf{C}_i \in H} \mathsf{C}_i \sqsubseteq \mathsf{O}\}$, and obviously $\mathcal{T} \cup \{\bigsqcap_{\mathsf{C}_i \in H} \mathsf{C}_i \sqsubseteq \mathsf{O}\} \cup \{\mathsf{C}^* \sqsubseteq \mathsf{C}_i \mid \mathsf{C}_i \in H\} \models \mathsf{C}^* \sqsubseteq \mathsf{O}$. Thus $\mathcal{T} \cup A \models \mathcal{O}$, A therefore solves **AP**. $\qquad\square$

In Colucci et al. (2003) the similar term *concept abduction* is used for a slightly different task: Given a L-knowledge base $\mathcal{T}$ and two L-concepts C and D that are satisfiable in $\mathcal{T}$, determine a L-concept H such that $\mathcal{T} \not\models \mathsf{C} \sqcap \mathsf{H} \sqsubseteq \bot$ and $\mathcal{T} \models \mathsf{C} \sqcap \mathsf{H} \sqsubseteq \mathsf{D}$. This problem can be reduced to subsumption-based abduction as shown in the following lemma:

Lemma 2.2 (Subsumption-based can simulate concept abd.)
Let $(\mathcal{T}, \mathsf{C}, \mathsf{D})$ *be a concept abduction problem over language L as defined by Colucci et al. (2003). Then, all solutions H to* $(\mathcal{T}, \mathsf{C}, \mathsf{D})$ *can be determined using subsumption-based abduction as defined in Definition 2.4.*

Proof. Define a subsumption-based abduction problem **SAP** for $(\mathcal{T}, \mathsf{C}, \mathsf{D})$ by $\mathbf{SAP}_{(\mathcal{T},\mathsf{C},\mathsf{D})} = (\mathcal{T}, \{\mathsf{C}\} \cup L^*(d), \mathsf{D})$, where $L^*(d)$ denotes the set of all semantically distinct L-concepts of role depth $\leq d$, with d defined as the maximum role depth occurring in C, D, or any axiom of $\mathcal{T}$. Further, let $Sol_{\mathbf{SAP}_{(\mathcal{T},\mathsf{C},\mathsf{D})}}$ be the solution set for the induced subsumption-based abduction problem, and $Sol_{\mathbf{SAP}_{(\mathcal{T},\mathsf{C},\mathsf{D})}|\mathsf{C}}$ its restrictions to subsets of $\mathcal{H}$ that do contain C.
Assume that H solves $\mathbf{SAP}_{(\mathcal{T},\mathsf{C},\mathsf{D})}$, and $H \in Sol_{\mathbf{SAP}_{(\mathcal{T},\mathsf{C},\mathsf{D})}|\mathcal{C}}$. Then $H \subseteq \mathcal{H}$, $\bigsqcap_{\mathsf{C}_i \in H} \mathsf{C}_i$ is consistent w.r.t. $\mathcal{T}$ (i.e. $\mathcal{T} \not\models \bigsqcap_{\mathsf{C}_i \in H} \mathsf{C}_i \sqsubseteq \bot$), and $\mathcal{T} \models \bigsqcap_{\mathsf{C}_i \in H} \mathsf{C}_i \sqsubseteq \mathsf{D}$. Now define $\mathsf{C_H} = \bigsqcap_{\mathsf{C}_i \in H} \mathsf{C}_i$. As $\mathsf{C} \in H$ by construction, it is also true that $\mathcal{T} \not\models \mathsf{C} \sqcap \mathsf{C_H} \sqsubseteq \bot$ and $\mathcal{T} \models \mathsf{C} \sqcap \mathsf{C_H} \sqsubseteq \mathsf{D}$, i.e. H solves the original concept abduction problem $(\mathcal{T}, \mathsf{C}, \mathsf{D})$. $\qquad\square$

The notion of abduction introduced in Definition 2.2 is therefore non-trivial and practically relevant, making it a good basis for our extension presented in Chapter 3.

2.2.3 A Critique of Logic-Based Abduction

A joint property of all formalizations of abduction introduced so far (independently of the concrete logic L used) is their requirement that a solution must completely explain the observation(s): There is no way of having a solution "explain the largest part of $\mathcal{O}$" or "make C almost subsume D". Typical application fields, such as media analysis (Castano et al., 2009) and industrial diagnostics (Hubauer et al., 2011b), are however characterized by an abundance of low-level observations due to a large number of sensors, whereas the knowledge base formalising the domain is often rough or incomplete since it must be created manually. As the following example shows, existing definitions of abduction are not sufficient in this context:

Example 2.1 (Diagnostics over incomplete domain formalisations)
The simple production system introduced before consists of a main control unit (MCU), a mechanical gripper and a conveyor which are both part of the transportation subsystem, and a PROFINET communication link to other systems of the factory. The system and its structure can be expressed using $\mathcal{EL}^+$ axioms as follows:

$$\mathsf{MCU} \sqsubseteq \exists\, \mathsf{partOf}\,.\, \mathsf{System}$$
$$\mathsf{Communications} \sqsubseteq \exists\, \mathsf{subsystemOf}\,.\, \mathsf{System}$$
$$\mathsf{Transportation} \sqsubseteq \exists\, \mathsf{subsystemOf}\,.\, \mathsf{System}$$
$$\mathsf{PROFINET} \sqsubseteq \exists\, \mathsf{belongsTo}\,.\, \mathsf{Communications}$$
$$\mathsf{Gripper} \sqsubseteq \exists\, \mathsf{belongsTo}\,.\, \mathsf{Transportation}$$
$$\mathsf{Conveyor} \sqsubseteq \exists\, \mathsf{belongsTo}\,.\, \mathsf{Transportation}$$
$$\mathsf{belongsTo} \circ \mathsf{subsystemOf} \sqsubseteq \mathsf{partOf}$$

The diagnostic expert knowledge of the system laid out before informally in Example 1.1 can then be formalized with the following set $\mathcal{T}$ of axioms:

$$\mathsf{MCU} \sqcap \exists\, \mathsf{partOf}\,.\, (\exists\, \mathsf{operatesIn}\,.\, \mathsf{PowerSupplyFluctuations})$$
$$\sqsubseteq \exists\, \mathsf{shows}\,.\, \mathsf{IntermittentOutages}$$
$$\mathsf{PROFINET} \sqcap \exists\, \mathsf{partOf}\,.\, (\exists\, \mathsf{operatesIn}\,.\, \mathsf{PowerSupplyFluctuations})$$
$$\sqsubseteq \exists\, \mathsf{shows}\,.\, \mathsf{SendingReceivingOK}$$
$$\mathsf{Gripper} \sqcap \exists\, \mathsf{partOf}\,.\, (\exists\, \mathsf{operatesIn}\,.\, \mathsf{PowerSupplyFluctuations})$$
$$\sqsubseteq \exists\, \mathsf{shows}\,.\, \mathsf{FullyFunctional}$$
$$\mathsf{MCU} \sqcap \exists\, \mathsf{partOf}\,.\, (\exists\, \mathsf{operatesIn}\,.\, \mathsf{ControlSWMalfunction})$$
$$\sqsubseteq \exists\, \mathsf{shows}\,.\, \mathsf{IntermittentOutages}$$

Conveyor $\sqcap \exists$ partOf . ($\exists$ operatesIn . ControlSWMalfunction)

$\sqsubseteq \exists$ shows . IrregularMovements

Gripper $\sqcap \exists$ partOf . ($\exists$ operatesIn . ControlSWMalfunction)

$\sqsubseteq \exists$ shows . IrregularMovements

The first axiom, for instance, expresses that a MCU (motor control unit) component being part of some (unspecified) larger system having a fluctuating power supply will be affected by showing intermittent outages. Similarly, the last axiom states that the mechanical gripper of a system affected by a control software malfunction will show irregularities in its movement patterns. We will return on the topic of how to model diagnostic knowledge later on in Section 3.1.

Let us now assume that based on the available sensor measurements, intermittent outages of the main control unit have been confirmed, and motion sensors signal flawless action of the gripper. Due to a general network problem throughout the factory, however, the working state of the PROFINET component cannot be asserted. These observations can be represented by the axiom set $\mathcal{O} = \{$MCU $\sqsubseteq \exists$ shows . IntermittentOutages, Gripper $\sqsubseteq \exists$ shows . FullyFunctional$\}$. In this situation, assuming fluctuations in the power supply as expressed by the assumption set $A_1 = \{$System $\sqsubseteq \exists$ operatesIn . PowerSupplyFluctuations$\}$ yields a valid solution (that is $\mathcal{T} \cup \{A_1\} \models \mathcal{O}$, the observations are logically entailed by the assumed diagnosis, given the knowledge base $\mathcal{T}$). Assuming a software malfunction, on the contrary, (i. e. $A_2 = \{$System $\sqsubseteq \exists$ operatesIn . ControlSWMalfunction$\}$) would not be considered a solution as it cannot account for the observation regarding the PROFINET system. Similarly, the explanation candidate assuming that both defects occur at the same time ($A_3 = A_1 \cup A_2$) would typically be rejected since it does not provide any additional expressive power over A_1 alone.

Extending on this situation, assume that a new vibration sensor is added to the system. The new sensor emits a signal whenever it detects vibrations that exceed a predefined threshold. It is not atypical in an application context that the diagnostic knowledge base is not updated instantly, possible reasons ranging from lack of information on how different faults affect vibrations to mundane lack of time. Continuing with the example, we now assume that the diagnostic unit is signalled the presence of significant low-frequency vibrations in addition to the previously discussed symptoms. This results in the extended observation set $\mathcal{O}' = \mathcal{O} \cup \{$MCU $\sqsubseteq \exists$ shows . LowFrequencyVibrations$\}$). As motivated before, naïve approaches to diagnostic or abductive reasoning would now try to explain the full observation set $\mathcal{O}$ '. Consequently, this would lead to *the invalidation of the previously determined solution A_1, as it cannot explain the observed vibrations (which are not even part of the diagnostic model). Moreover, unless the knowledge bases are updated accordingly, the system may fail to find any solution at all. This illustrates the counterintuitive effect that an additional (and potentially completely unrelated) piece of information may cause the diagnostic problem to fail completely.[7]*

This behaviour severely hinders the practical applicability of logic-based abduction to real-world industrial applications, where an ever-growing amount of sensor data almost inevitably generates pieces of information that the current representation of the domain knowledge cannot account for. We therefore suggest that the classic definition of logic-based abduction is too strict for information-intensive applications, leading to overly complex diagnoses or even the failure to produce any solution at all in the presence of such spurious observations. This is justified by revisiting the requirements defined in Section 1.2, where logic-based abduction alone fails to meet Requirements **R3** and **R5**. For very simple knowledge bases, a remedy could be to identify and remove problematic observations in a preprocessing step, resulting in an approach that meets Requirements **R1** to **R3**, and the restriction of Requirement **R4** to classical solutions (but nevertheless fails w. r. t. Requirement **R5**). This is however not feasible for reasonably complex formalisations since the (ir-)relevance of a piece of information depends on the analysis result and is thus not known beforehand.

A more general solution can be found in the definition of Scott-entailment relations (Scott, 1974) and derived approaches such as the notion of causal abduction presented by Bochman (2007), where an axiom set is considered as an explanation for a set of observations if it entails at least one of the observations[8]. However, the very general framework laid out by Bochman does not define an order over the set of observations, and thus permits no detailed ranking over the typically large set of possible explanations. More recently, Kaya (2011); Peraldi (2011) proposed a different approach

---

[7]As can easily be seen, even this simple scenario makes use of relational structures both for describing the partonomy of the system and for defining how faults in the system affect its observable state. While for a specific case it would obviously be possible as well to use a simple propositional logic knowledge base to represent the same information, the availability of relations and existential quantification makes the representation much more concise and intelligible. This is a vital factor to ensure that experts can work efficiently with the domain model and update it as needed without getting lost in hundreds of propositional rules.

[8]More formally, Scott-entailment between two axiom sets X and Y holds if X entails some subset of Y, i. e. $Cn(X) \cup Y \neq \emptyset$ with $Cn(X)$ denoting the deductive closure of the axiom set X.

to address the problem of inexplicable observations in an automated way: any observation that cannot be explained from the formal representation is simply assumed to be true. Although this solution has been shown to provide reasonable results e. g. in the context of media interpretation, the equalisation of assumptions and observations leads to a loss of information and subsequently to the loss of other potentially relevant solutions). Extending on this research, Nafissi (2013) employs a beam-search approach to control the generation of solution candidates in an abductive Markov Logic setup. However, while the authors can significantly reduce the size of solution space that is explored using a heuristic approach, they do not explicitly address the trade-off between expressiveness and simplicity of a single solution (instead, they use one-dimensional quality score). The variant of logic-based abduction proposed in this thesis can be understood as a natural continuation of this line of research: It provides robustness w. r. t. deficient domain representations, but does neither require a tradeoff between expressiveness and simplicity, nor assume their separability. We present our solution in Chapter 3, but beforehand we introduce basics from order and hypergraph theory required for a clear definition of both the problem and our solution to it.

We conclude this section with a short explanation why we represent observations by means of terminological axioms in Example 2.1. Firstly, note that this representation is not required by the framework itself, but only for a specific instantiation for the description logic $\mathcal{EL}^+$ which we will introduce in Section 3.3.2 (a restriction which can however be lifted by the introduction of nominals or enumerated classes, as explicated in Section 3.4.1). Yet, to keep the running example consistent throughout this thesis, we use a terminology-based notation for observations right from the start.

2.3 Orders

By a *preorder* $\preceq$ on some set X we understand as usual a binary relation $\preceq \subseteq X \times X$ that is reflexive, i. e., for all $x \in X : x \preceq x$, and transitive, i. e., $\forall x, y, z \in X$: if $(x \preceq y)$ and $(y \preceq z)$, then $(x \preceq z)$. If we want to make explicit the set X over which a relation $\preceq$ is defined, we write $\preceq_X$. Two additional important orders can be derived from $\preceq$: $x \simeq y$ (equivalence) is a shorthand for $x \preceq$ and $y \preceq x$ being both valid, whereas $x \prec y$ (strict preorder) is used as an abbreviation for $x \preceq y$ and it is not the case that $y \preceq x$. Two elements $x, y \in X$ are called incomparable w. r. t. to $\preceq_X$ if any only if neither $x \preceq_X y$ nor $y \preceq_X x$. The preorder $\preceq_X$ is called total if no two elements of its domain X are incomparable, otherwise $\preceq_X$ is a partial preorder. An element $x \in X$ is minimal w.r.t. $\preceq_X$, or an $\preceq_X$-*minimal* element, if and only if there is no $y \in X$ such that $y \prec_X x$. More generally, $x \in X'$ is a $\preceq_X$ minimal element of $X' \subseteq X$ if there is no $y \in X'$ with $y \prec_X x$.

A (partial or total) *order* is a (partial or total) preorder that additionally fulfils the antisymmetry property, i.e., $\forall x, y \in X$: if $x \preceq y$ and $y \preceq x$, then $x = y$. Note that many "natural" preference relations (such as $\leq$ over set or real numbers) are antisymmetric and therefore orders. However, there exist interesting preference relations over sets of axioms (as needed for logic-based diagnosis) which are not antisymmetric (for instance, two sets X, Y containing the same number of axioms need not be identical although). As this thesis aims at defining a general framework first, and only study specific instantiations and their properties after that, the more general setting of preorders is of special interest.

Given any binary relations R_X, R_Y (not necessarily preorders) over the sets X, Y respectively, one can define the Cartesian product $R_{X \times Y} \subseteq R_X \times R_Y = (X \times Y) \times (X \times Y)$ as usual, letting

$$(x_1, y_1) \, R_{X \times Y}(x_2, y_2) \quad \text{iff} \quad x_1 R_X x_2 \text{ and } y_1 \, R_Y \, y_2$$

It is a well known fact from model theory that properties that can be represented as universal Horn formulas are preserved under the Cartesian product constructions. Hence, as reflexivity, transitivity, and antisymmetry are all universal Horn formulas, we may restate the following fact that the Cartesian product of preorders (partial orders) is again a preorder (partial) order:

Lemma 2.3 (Cross product of (pre-)orders)
If $\preceq_X, \preceq_Y$ are preorders (partial orders), then $\preceq_X \times \preceq_Y$ is a preorder (partial order), too.

As the Cartesian product of preorders is again a preorder, it gives rise to the notion of minimal elements $(x, y) \in X \times Y$ over the cross-product of both sets. Considering the preorders as preference relations on equally relevant dimensions X and Y, it is a natural assumption to think of the minimal pairs over the pairs on the Cartesian product as the most preferable ones. This idea can be formally grounded within the notion of Pareto-optimality from the theory of economics. Intuitively, the so-called weak definition of Pareto-optimality states that trying to make one component of a Pareto-optimal element strictly better must lead to one component being not better

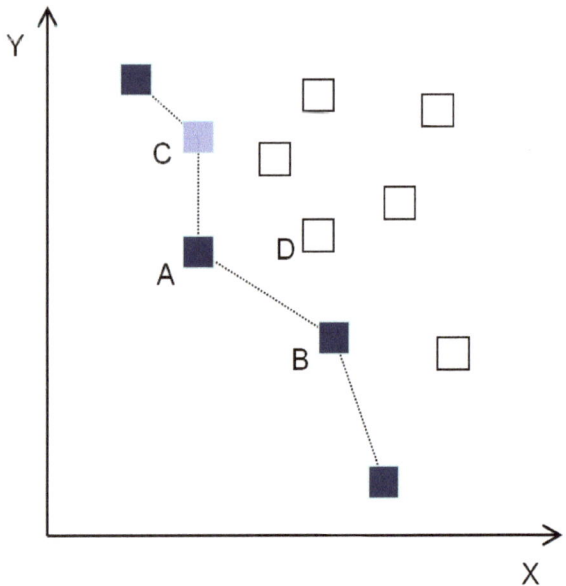

Figure 2.1: Pareto-optimality and Dominance

(i. e. either incomparable or worse), the strong notion demands a component being strictly better. Figure 2.1 visualizes the different notions for total orders. In this example, the points A and B are both weakly and strongly Pareto-optimal, whereas C is weakly Pareto-optimal, but not strongly. Point D is said to be *dominated* by A (A is smaller than D in all dimensions). In the next definition, we formalize these intuitions mathematically.

Definition 2.5 (Pareto-optimality)
Let $\overrightarrow{X} = X_1 \times \ldots \times X_n$ be the Cartesian of the sets X_i and $\preceq_i$ be preorders on X_i. Then $\vec{x} = (x_1, \ldots, x_n) \in X$ is said to be weakly Pareto-optimal for X iff:
For all $\vec{x}' \in \overrightarrow{X}$: If there is some i with $1 \leq i \leq n$ s.t. $x_i' \prec_i x_i$, then there is a j with $1 \leq j \leq n$ such that not $x_j' \preceq x_j$.

  Moreover, $\vec{x} = (x_1, \ldots, x_n) \in \overrightarrow{X}$ is said to be strongly Pareto-optimal for X iff:
For all $\vec{x}' \in \overrightarrow{X}$: If there is some i with $1 \leq i \leq n$ s.t. $x_i' \prec_i x_i$, then there is a j with $1 \leq j \leq n$ such that $x_j \prec x_j'$.

As we show next, minimal and weakly Pareto-optimal elements coincide for the Cartesian product construction introduced before:

Lemma 2.4 (Weakly Pareto-optimal elements)
The weakly Pareto-optimal elements over the Cartesian product of preorders relations are exactly the minimal elements.

Proof. Let $\preceq_X$ denote the Cartesian product of component preorders $\preceq_{X_i}$ ($1 \leq i \leq n$) from Definition 2.5. Let be $\vec{x}$ be a (weakly) Pareto-optimal element. Assume for contradiction that there is a $\vec{x}' \prec_X \vec{x}$. That means that that for all i one has $x_i' \preceq x_i$ and at least for one j $x_j' \prec_j x_j$, which contradicts the Pareto-optimality of $\vec{x}$. To prove the other direction assume that $\vec{x}$ is a minimal element. Let $\vec{x}'$ be such that there is some j with $x_j' \prec_j x_j$. If for all other i we had $x_i' \preceq_i x_i$, then we would have $\vec{x}' \prec \vec{x}$ contradicting the minimality of $\vec{x}$. $\qquad\square$

If the component preorders also fulfil the property of connectedness or totality (i.e., for all x, x' either $x \preceq x'$ or $x' \preceq x$ or $x = x'$ holds), then the notions of weak Pareto-optimality and strong Pareto-optimality collapse into one, as $\neg(x \preceq x')$ entails $x' \prec x$. So we get:

Lemma 2.5 (Strongly Pareto-optimal elements)
The minimal elements of Cartesian products of total preorders are just the (strongly, weakly) Pareto-optimal elements.

Total preorders are an important structure in belief revision and non-monotonic reasoning (for details see Booth & Meyer, 2011).

Set inclusion $\subseteq$ over the power set $\mathcal{P}(X) = \{Y \mid Y \subseteq X\}$ of some set X is a special partial order which will be relevant for abduction. Then, by Lemma 2.3, $\preceq \subseteq \mathcal{P}(X) \times \mathcal{P}(X)$ is a preorder on the Cartesian product of $\mathcal{P}(X)$ with itself. This preorder $\preceq$ is said to be *monotonic (anti-monotonic) for set inclusion* if and only if for all $x, x' \in \mathcal{P}(X)$ with $x \subseteq x'$ one also has $x \preceq x'$ ($x' \preceq x$).

2.4 Hypergraphs

A *graph* is a mathematical structure that connects pairs of *vertices* by means of (directed or undirected) *edges*. *Hypergraphs* generalise graphs by extending the definition of an edge from a binary to an n-ary relation. This way, any (non-empty) subset of vertices can form an edge that may again be undirected or directed. In the following, we introduce key notions about hypergraphs which are needed for our work. Additional information can be found, among others, in the work by Nielsen (2001), whose notation is used here in adapted form.

Definition 2.6 (Directed hypergraph)
A directed hypergraph is a tuple $\mathcal{H} = (V, E)$ where V is a finite, nonempty set of vertices, and $E \subseteq (\mathcal{P}(V) \setminus \emptyset) \times V$ is a finite set of edges. For each edge $e = (T, h) \in E$, $T(e)$ denotes the tail *of e, and $h(e)$ denotes its* head.

$\mathcal{H} = (V, E)$ is a weighted directed hypergraph if V is defined like before, and $E \subseteq \mathcal{P}(V) \times V \times W$ for an arbitrary space W of weights. The weight of an edge $e = (T, h, w)$ is denoted by $w(e)$.

Weights in graphs or hypergraphs are typically used to represent some evaluation of the edge in terms of quality, cost, or the like. It is typically desirable to extend this evaluation from single edges to sets of edges. To this end, we need an operator for combining weights; the mathematical structure of a monoid captures basic requirements for such a combination:

Definition 2.7 (Monoid)
Let S be a set and $\otimes$ a binary operation. $(S, \otimes)$ is a monoid if and only if

    a) $\forall s_1, s_2 \in S : s_1 \otimes s_2 \in S$ (S is closed under $\otimes$)

    b) $\forall s_1, s_2, s_3 \in S : (s_1 \otimes s_2) \otimes s_3 = s_1 \otimes (s_2 \otimes s_3)$ ($\otimes$ is associative)

    c) $\exists e \in S \forall s \in S : e \otimes s = s = s \otimes e$ (existence of a neutral element)

$(S, \otimes)$ is a commutative monoid if and only if it is a monoid and

    d) $\forall s_1, s_2 \in S : s_1 \otimes s_2 = s_2 \otimes s_1$ ($\otimes$ is commutative)

In standard graphs, a path (of length n) is a sequence $e_1, \ldots, e_i, e_{i+1}, \ldots, e_n$ of edges such that two consecutive edges share a node, for example $e_i = (v_i, v_{i+1}), e_{i+1} = (v_{i+1}, v_{i+2})$, and so on. If the graph is weighted (that is, the edges each have an attributed weight), the edge weights along the path can be aggregated to yield the path weight. Reflecting the generalization of edges to arbitrary sets of vertices, a hyperpath represents a consecutive chain of connections between two sets of vertices. Intuitively, there is a hyperpath from S (start vertices) to g (goal vertex) if there is a hyperedge connecting some intermediate set of vertices Y to g, and each $y_i \in Y$ is in turn reachable from S via a hyperpath. Definition 2.8 formalizes this intuition.

Definition 2.8 (Directed hyperpath)

Let $\mathcal{H} = (V, E)$ be a directed hypergraph. Then $p_{S,g} = (V_{S,g}, E_{S,g})$ is a simple directed hyperpath in $\mathcal{H}$ from S to g if and only if

    a) $g \in S$ and $p_{S,g} = (\{g\}, \emptyset)$, or

    b) $\exists e \in E : h(e) = g \wedge T(e) = \{y_1, \ldots, y_k\} \wedge \forall 1 \leq i \leq k \; \exists p_{S,y_i} :$
$$\left(V \supseteq V_{S,g} = \{g\} \cup \bigcup_{y_i \in T(e)} V_{S,y_i} \wedge E \supseteq E_{S,g} = \{e\} \cup \bigcup_{y_i \in T(e)} E_{S,y_i} \right)$$

Moreover, $p_{S,G} = (V_{S,G}, E_{S,G})$ is a directed hyperpath in $\mathcal{H}$ from S to G if and only if

    c) $\forall g \in g : p_{S,g} = (V_{S,g}, E_{S,g})$ is a directed hyperpath in $\mathcal{H}$ from S to g,
        $V_{S,G} = \bigcup_{g \in G} V_{S,g}$ and $E_{S,G} = \bigcup_{g \in G} E_{S,g}$

If $\mathcal{H}$ is a weighted hypergraph, let W denote its weight space and let $\otimes$ be a binary operation such that $(W, \otimes)$ is a commutative monoid called the weight system *of $\mathcal{H}$. The notion of a* weight *then extends from hyperedges e to simple hyperpaths $p_{S,g} = (V_{S,g}, E_{S,g})$ by $w(p_{S,g}) = \bigotimes_{e \in E_{S,g}} w(e)$, and to hyperpaths $p_{S,G} = (V_{S,G}, E_{S,G})$ by $w(p_{S,G}) = \bigotimes_{e \in E_{S,G}} w(e)$.*

It is easy to see that a simple hyperpath is thus a sub-hypergraph of $\mathcal{H}$ connecting (a subset of) the vertices in S to the vertex g. Similarly, a hyperpath from S to G is a sub-hypergraph of $\mathcal{H}$ that connects (a part of) the vertices in S to all vertices in G. The cost of a hyperpath is determined by "adding up" the costs of its hyperedges. The requirement of $(W, \otimes)$ being a commutative monoid is derived from the frequently-used weight system $(\mathbb{N}, +)$. While other systems such as $(\mathbb{R}, +)$ correspond to algebraic structures that extend commutative monoids with additional properties (such as the existence of inverses), these additional properties are not relevant for our work.

In this chapter, we have provided the tools needed to give a clear, formal definition of relaxed abduction in Section 3.2. In a nutshell, we will define solutions to a relaxed abduction problem to be the Pareto-optimal elements of a solution space. Then, we will show how derivations of an axiom in a so-called proof tree correspond to directed weighted hyperpaths, and employ this finding to devise an algorithm for computing the set of solutions of a relaxed abduction problem as shortest hyperpaths in the proof tree.

3 Relaxed Abduction

In this chapter we introduce relaxed abduction, a novel inference method designed specifically for the requirements posed by flexible information interpretation as presented in Section 1.2. As a starting point, Section 3.1 presents a formal framework for diagnostics of technical systems based on established ISO-standardised terminology, providing a basis for the examples used throughout this chapter. Next, Section 3.2 addresses the main theoretical contributions of this chapter by motivating and formalising relaxed abduction, analysing its relation to standard abduction, and presenting several concrete instantiations of the general framework including a mapping to the diagnostic tasks introduced in Section 3.1. In Section 3.3 we then investigate practical algorithms for solving relaxed abduction in the context of description logics. We present both a naïve but generic solution, and a more sophisticated algorithm tailored specifically to the $\mathcal{EL}^+$ formalism that is used for representing diagnostic ontologies. Early pruning of suboptimal solution candidates is investigated as a means of increasing the efficiency of the $\mathcal{EL}^+$ algorithm. After giving details on how the basic framework can be extended to support, for instance, the efficient addition and retraction of observations in Section 3.4, we conclude this chapter in Section 3.5 by relating relaxed abduction to other approaches to information interpretation.

3.1 Formalising Diagnostics using Description Logics

A lighthouse task for information interpretation in industrial contexts is diagnostics of technical systems, which plays a crucial role in the operation of modern industrial machinery. We have studied requirements and solutions related to such use cases in depth in Hubauer et al. (2011b) and Legat et al. (2011). For relating the features provided by relaxed abduction to concrete applications, we introduce in this section a conceptual model for diagnostics, and relate several of its instantiations to concrete relaxed abductive reasoning tasks.

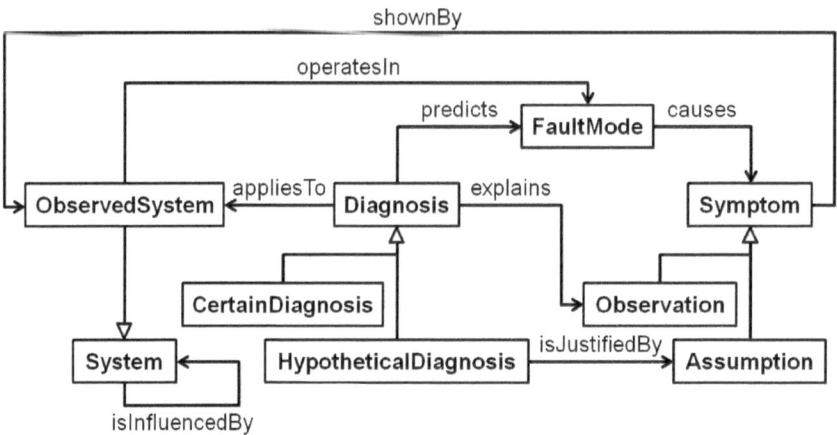

Figure 3.1: A conceptual model for diagnostics

3.1.1 A Conceptual Model for Diagnostics

Diagnostic processes typically aim at finding the root cause(s) for a system operating in a fault mode and thus showing certain symptoms of abnormal behaviour. Automating these processes can support the operator in identifying faults and thereby reduce costly system downtimes. The standard ISO 13379 on "Condition Monitoring and Diagnostics of Machines" (ISO, 2003) set by the International Organization for Standardization lays a common basis for diagnostics of typical technical equipment in industry. Building upon on the terminology used in this norm, we have extracted a conceptual model to characterize the type of diagnostics to be supported. It is depicted in Figure 3.1 in UML notation and explained in the following, reflecting several aspects found in ISO 13379.

One particular aspect stressed in ISO 13379 is the structural decomposition of systems, and the influence that (sub-)systems can have on each other with respect to fault propagation. An *observed system* to be diagnosed typically stands in relation to surrounding *systems* by which it is *influenced*; examples of influential relationships are sub-part composition and electrical wiring. Another aspect of ISO 13379 is the connection between fault modes as a cause and symptoms as an effect thereof: In case of diagnostics of abnormal behaviour, the observed system *operates in* a certain *fault mode*, which *causes* the influenced systems (e. g. the parts of the observed system) to *show* specific *symptoms*. Finally, a diagnostics process according to ISO 13379 is triggered by detection of an anomaly in the system behaviour, and yields as a result a *diagnosis* that *predicts* a particular fault mode for the observed system. Due to the causal relationship between fault modes and symptoms, a diagnosis explains symptoms that play the role of *observations*. Moreover, a diagnosis can be either *certain* or *hypothetical*, in which case it *is justified by* additional symptoms that have not been observed but play the role of *assumptions*.

The knowledge supporting machine diagnostics can be formalised in two different ways. Firstly, diagnostic knowledge can be engineered with a deep understanding of the underlying machinery from first principles, typically by the engineers who constructed the machine. This way the causal relationship between fault modes and symptoms is well represented, allowing for the derivation of resulting symptoms from a given fault. Alternatively, diagnostics knowledge can be designed by observation, looking at the system as a black box, for instance by the engineers using the machine. The modelled relationship between fault modes and symptoms is then considered in the reverse direction, allowing for the derivation of faults from symptoms. In analogy, we follow Brazier et al. (1996) in making the distinction between a *causal* and an *anti-causal* way of modelling knowledge for diagnostics. A similar distinction can be found in the work by Poole (1988), which analyses the commonalities and differences of consistency-based, abductive and rule-based diagnosis, and characterises them as specific instantiations of the Theorist framework. As explicated by Poole, consistency-based diagnosis can make use of both causal models of correct behaviour and anti-causal fault models to derive hypotheses on the set of abnormal components, given descriptions of normal and abnormal behaviour observed. At the same place, rule-based diagnostics is defined to use non-causal domain models, whereas abductive reasoning is connected to causal representations of the domain.

In the chosen context of logic-based approaches for diagnostics, a causal diagnostic knowledge base comprises implications of the form "Fault implies Symptom" which support a deduction mechanism in deriving symptoms whereas in the anti-causal case, we have implications of the form "Symptom implies Fault" that support deductive derivation of faults. The following subsection investigates the use of $\mathcal{EL}^+$ for modelling causal as well as anti-causal diagnostic knowledge.

3.1.2 Formalising Diagnostics in $\mathcal{EL}^+$

As pointed out before, the description logic $\mathcal{EL}^+$ can be used for modelling diagnostic information. Based on this observation and the fact that our

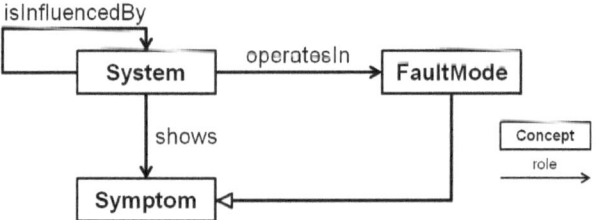

Figure 3.2: ER representation of core entities in diagnostics

current implementation of consequence-driven relaxed abduction fully implements the features of this DL, we have formalized parts of the abstract model for diagnostics introduced before in a set of ontologies and a mapping of diagnostics tasks to reasoning problems in the $\mathcal{EL}^+$ formalism: The core entities involved in diagnostics are captured in the knowledge base $\mathcal{M}_{sys}$ (core system model, depicted in Figure 3.2) that serves as a backbone ontology for diagnostics tasks. Note that only a part of the elements from the conceptual model shown in Figure 3.1 are represented explicitly in the knowledge base $\mathcal{M}_{sys}$. Extending on the original ISO definition, however, we have added a subclass-relation between FaultMode and Symptom. The intuition behind this decision is to facilitate hierarchical diagnostics scenarios where a diagnosis referring to one component can straightforwardly be used as a symptom of the complete system. To this ends, the role inclusion axiom

$$\text{isInfluencedBy} \circ \text{operatesIn} \sqsubseteq \text{shows}$$

(not visible in the graphical depiction) has been added to $\mathcal{M}_{sys}$ as well.

In a concrete application the targeted systems are modelled in a separate ontology $\mathcal{M}_{cust}$ (for custom model) that extends on the entities defined in $\mathcal{M}_{sys}$ and defines the static system knowledge for the diagnostic reasoning task. The dynamic knowledge is captured in additional ontologies described in the following paragraphs and depicted in Figure 3.3 along with the dependencies between the ontologies.

Representing Diagnostic Knowledge

Causal Modelling. Diagnostic knowledge represented causally is captured in an ontology $\mathcal{M}_{caus}$ of axioms that have the form

$$\langle \text{Sys}_i \rangle \sqcap \exists \langle \text{Rinf}_i \rangle \,.\, (\exists\, \text{operatesIn} \,.\, \langle \text{Fault} \rangle) \sqsubseteq \textstyle\bigsqcap_j \exists\, \text{shows} \,.\, \langle \text{Sym}_{i,j} \rangle \,,$$

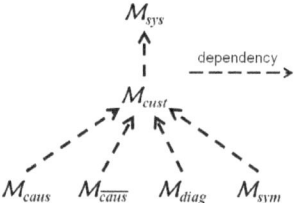

Figure 3.3: A hierarchy of diagnostic ontologies

where $\langle C \rangle$, $\langle r \rangle$ denote variables that range over concepts and roles respectively, $\langle Sys_i \rangle$ is some system, $\langle Sym_{i,j} \rangle$ are concrete symptoms, and $\langle Rinf_i \rangle$ is a relation expressing that one system influences another. We moreover require that $\mathcal{M}_{sys} \cup \mathcal{M}_{cust} \models \{\langle Fault \rangle \sqsubseteq FaultMode, \langle Sym_{i,j} \rangle \sqsubseteq Symptom, \langle Rinf_i \rangle \sqsubseteq$ isInfluencedBy$\}$. These axioms state that a system operating in fault mode $\langle Fault \rangle$ manifests itself by the influenced systems $\langle Sys_i \rangle$ (e. g. its subsystems) showing certain symptoms $\langle Sym_{i,j} \rangle$. Note that the effects of a certain fault $\langle Fault \rangle$ are represented by several axioms, one per influenced component $\langle Sys_i \rangle$.[1]

Anti-Causal Modelling. Diagnostic knowledge modeled anti-causally is captured in an ontology $\mathcal{M}_{\overline{caus}}$ of axioms that have the form

$$\textstyle\bigsqcap_j \exists \, shows \, . \, \langle Sym_{i,j} \rangle \sqsubseteq \langle Sys_i \rangle \sqcap \exists \, \langle Rinf_i \rangle \, . \, (\exists \, operatesIn \, . \, \langle Fault \rangle) \, ,$$

where $\langle C \rangle$, $\langle r \rangle$ denote variables that range over concepts and roles respectively, $\langle Sys_i \rangle$ is some system, $\langle Sym_{i,j} \rangle$ are concrete symptoms, and $\langle Rinf_i \rangle$ is a relation expressing that one system influences another. We moreover require that $\mathcal{M}_{sys} \cup \mathcal{M}_{cust} \models \{\langle Fault \rangle \sqsubseteq FaultMode, \langle Sym_{i,j} \rangle \sqsubseteq Symptom, \langle Rinf_i \rangle \sqsubseteq$ isInfluencedBy$\}$. These axioms state that a system that influences other systems $\langle Sys_i \rangle$ showing certain symptoms $\langle Sym_{i,j} \rangle$ is operating in a particular fault mode $\langle Fault \rangle$. Note that the difference between causal and anti-causal modelling in our $\mathcal{EL}^+$ formalization lies in the direction of the subsumption symbol.

Biconditionals. As pointed out for instance in Kowalski (2011), causal representations are considered a more natural way of modelling but require

---

[1]The example used throughout this thesis which was introduced in Example 2.1 is based on such a causal diagnostic model. The structural description of the system represented in $\mathcal{M}_{sys}$, however, has been simplified for the ease of presentation.

the use of abductive inference for realizing diagnostics, whereas anti-causal representations allow for the use of deduction to realize (some kind of) diagnostic reasoning. One solution to this problem is the use of biconditionals, where equivalence is used instead of implication (or, in the context of DLs $\equiv$ instead of $\sqsubseteq$ or $\sqsupseteq$). Using a process known as Clark's Completion (for details see Clark, 1977), such a biconditional representation can be generated automatically from a (natural) causal one, which can be understood as employing a local closed world assumption on the diagnostic knowledge (Kowalski, 2011). Whereas this process allows for a domain model to be interpreted in the reverse direction, it still resorts to (sound, complete, and monotonic) deduction for this evaluation. The problem of missing observational data is however not solved. Therefore, the algorithms proposed later do *not* employ Clark's Completion.

Representing Symptoms

The concept Symptom is an element of the signature of $\mathcal{M}_{sys}$, while specific symptoms $\langle \mathsf{Sym}_{i,j} \rangle$ are modelled as subconcepts of Symptom in $\mathcal{M}_{cust}$ such that $\mathcal{M}_{sys} \cup \mathcal{M}_{cust} \models \langle \mathsf{Sym}_{i,j} \rangle \sqsubseteq \mathsf{Symptom}$. All symptom observations on (parts of) systems described in $\mathcal{M}_{cust}$ are captured by axioms $\langle \mathsf{Sys}_i \rangle \sqsubseteq \exists \mathsf{shows} . \langle \mathsf{Sym}_{i,j} \rangle$ in $\mathcal{M}_{sym}$.

Representing Fault Modes

Similarly, the concept FaultMode is an element of the signature of $\mathcal{M}_{sys}$ that serves as a superclass for any specific system fault mode modelled as a concept $\langle \mathsf{Fault} \rangle$ in $\mathcal{M}_{cust}$ such that $\mathcal{M}_{sys} \cup \mathcal{M}_{cust} \models \langle \mathsf{Fault} \rangle \sqsubseteq \mathsf{FaultMode}$.

Representing Diagnoses

A diagnosis candidate is an axiom of the form $\langle \mathsf{Sys}_i \rangle \sqsubseteq \exists \mathsf{operatesIn} . \langle \mathsf{Fault} \rangle$ such that $\mathcal{M}_{sys} \cup \mathcal{M}_{cust} \models \{\langle \mathsf{Sys}_i \rangle \sqsubseteq \mathsf{System}, \langle \mathsf{Fault} \rangle \sqsubseteq \mathsf{FaultMode}\}$ (also called a hypothesis). All potential diagnoses for the system(s) described in $\mathcal{M}_{cust}$ are captured in the ontology $\mathcal{M}_{diag}$.

Having introduced our meta model for representing diagnostic information using description logics in this section, we now turn towards the formalisation of relaxed abduction. At the end of the following section, we will return to the topic of diagnostics introduced here, and show how relaxed abduction can be applied to solving diagnostic problems.

3.2 Formalization and Properties of Relaxed Abduction

Relaxed Abduction is a knowledge-based reasoning mechanism that makes logic-based abduction robust to deficiencies in the underlying knowledge base. In Section 2.2.3 we have shown how the typical definition of logic-based abduction requires all observations to be entailed by a valid solution, which may lead to the generation of overly complex solutions (or the failure to find any) in the presence of observations not foreseen at modelling time. In contrast to that, relaxed abduction drops the need to explain every single observation, replacing it by the objective of finding solutions that combine maximum explanatory power with minimum need for assumptions. Hobbs et al. strikingly state this principle as "[we] want to get more bang for the buck" in their seminal work on abductive text interpretation (see Hobbs et al., 1993), and point out the close relation to the principles of *simplicity* and *consilience*. Following Thagard (1978), consilience measures how much a description explains, making the explanatory power of competing explanations measurable. In this context, simplicity can then be seen as a restricting factor to consilience: As almost any observation can be explained by making a hypothesis arbitrarily complex, simplicity provides a measure of how efficiently this expressive power is realized.

3.2.1 Bi-Criteriality of Information Interpretation

This antithetic character of simplicity and consilience is conceptually quite similar to the relationship between precision and recall in information retrieval, where alternative answer sets are typically weighed against each other based on a linear and complete order derived from precision and recall. This reduction is frequently accomplished by means of a total mapping $F : [0; 1] \times [0; 1] \mapsto [0; 1]$ called *F-measure* taken from a parametrized family of functions defined by

$$F_\beta(precision, recall) = \frac{(1 + \beta^2) \cdot precision \cdot recall}{(\beta^2 \cdot precision + recall)}.$$

This approach has been proven highly valuable in the field of information interpretation as a means of making solutions easily comparable.

It is therefore a straightforward idea to linearise the preorders induced by simplicity and consilience in a similar way for information interpretation. However, such an approach requires an a priori decision on the "exchange rate" between simplicity and consilience as represented by the parameter β in the equation above. Depending on the concrete choice of this parameter, relevant solutions may then get lost due to this tradeoff. The approach taken in Hobbs et al. (1993), called weighted abduction, is similar in this respect and therefore shares this problem: The authors define real-valued costs for assumptions and for unexplained observations, which can equivalently be seen as considering all observations abducible, and assuming them if no explanation can be found. Costs of a solution are then determined by summing up the costs of all assumptions.

As has been criticized before (see Norvig & Wilensky, 1990), weighted abduction (and, more generally, any approach based on such a dimensionality reduction) tacitly presupposes the *commensurability* of the qualities involved. As Norvig & Wilensky illustrate, this requirement is not met in the context of abduction for several reasons: Most importantly, such a reductionist approach makes it impossible to distinguish explanations with little expressive power (i. e. non-consilient ones) from those requiring a large number of assumptions (thereby violating simplicity). Obviously, neither "magical" solutions explaining all observations at the cost of highly implausible assumptions nor trivial solutions where an observation is explained matter-of-factly by simply assuming it are of much use in real-world applications and should therefore be clearly distinguishable from other more useful results. This shows that bi-criteriality is an inherent property of information interpretation which should be handled in an explicit way – as Norvig & Wilensky put it, "a single number cannot represent (...) the cost and quality of an explanation". The next section introduces a relaxation of the well-known framework of logic-based abduction discussed in Section 2.2 that addresses this challenge in a formally well-defined way.

3.2.2 Relaxing Logic-Based Abduction

Relaxed abduction explicitly addresses the bi-criterial nature of abduction as to provide a flexible mechanism that is robust to imperfections of the underlying knowledge base. Intuitively, this is achieved by alleviating the requirement that a solution must necessarily explain every single observation made. Differently from existing work, relaxed abduction refrains from weighing assumptions and observations against each other. Instead, solutions to a relaxed abduction problem explicitly include information on both assumptions taken and observations explained; simplicity and consilience of these solutions are then enforced by employing two (pre)order relations. This intuition is formalized in the following definition:

Definition 3.1 (Relaxed abduction problem)

A relaxed abduction problem is a 5-tuple $\mathbf{RAP} = (\mathcal{T}, \mathcal{A}, \mathcal{O}, \preceq_{\mathcal{A}}, \preceq_{\mathcal{O}})^2$, *where*

- $\mathcal{T}$ *is a set of $L_{\mathcal{T}}$-axioms that formalise the domain,*
- $\mathcal{A}$ *is a set of $L_{\mathcal{A}}$-axioms representing assumptions,*
- $\mathcal{O}$ *is a set of $L_{\mathcal{O}}$-axioms representing observations,*
- $\preceq_{\mathcal{A}} \subseteq \mathcal{P}(\mathcal{A}) \times \mathcal{P}(\mathcal{A})$ *is a preorder over sets of assumptions that is monotonic for set inclusion, and*
- $\preceq_{\mathcal{O}} \subseteq \mathcal{P}(\mathcal{O}) \times \mathcal{P}(\mathcal{O})$ *is a preorder over sets of observations that is anti-monotonic for set inclusion.*

A pre-solution to $\mathbf{RAP}$ *is a pair* $(A, O) \in \mathcal{P}(\mathcal{A}) \times \mathcal{P}(\mathcal{O})$ *such that* $\mathcal{T} \cup A$ *is consistent and* $\mathcal{T} \cup A \models O^3$. *A solution is an $\preceq$-minimal element of the set of pre-solutions, where* $\preceq \, = \, \preceq_{\mathcal{A}} \times \preceq_{\mathcal{O}}$ *is the Cartesian product of the original preference relations. We write* $(A, O) \models \mathbf{RAP}$ *to state that* (A, O) *solves* $\mathbf{RAP}$. *The solution set for* $\mathbf{RAP}$ *is defined by* $Sol_{\mathbf{RAP}} := \{(A, O) \mid (A, O) \models \mathbf{RAP}\}$.

To ensure decidability of the $\mathbf{RAP}$, *we require that the logic* $L := L_{\mathcal{T}} \cup L_{\mathcal{A}} \cup L_{\mathcal{O}}$ *that unifies the representation languages is decidable (this also includes the special case where* $L = L_{\mathcal{T}} = L_{\mathcal{A}} = L_{\mathcal{O}}$). *We use the notion of an* L-$\mathbf{RAP}$ *if the specific language* L *is important and not clear from the context.*

As can easily be seen, $\preceq$ is a preorder by Lemma 2.3. The intuition of $\preceq_{\mathcal{A}}$ and $\preceq_{\mathcal{O}}$ is to capture the notions of simplicity and consilience, respectively. The (anti-) monotonicity requirements guarantee these properties on a simple per-element basis without taking the logical structure connecting the axioms into account. This can be accomplished by enforcing stricter adequateness properties on the preorders, namely irredundancy (defined analogously to Eiter & Gottlob, 1995, Def 2.4) and its dual:

Definition 3.2 (Irredundancy, dual irredundancy)

Let $\mathbf{RAP} = (\mathcal{T}, \mathcal{A}, \mathcal{O}, \preceq_{\mathcal{A}}, \preceq_{\mathcal{O}})$. *The preorder* $\preceq_{\mathcal{A}}$ *is called* irredundant *if and only if for all pre-solutions* $(A, O), (A', O')$ *to* $\mathbf{RAP}$: *If* $A \subsetneq A'$, *then*

---

[2]Actually, it would be more in line with terminology from e.g. automata theory to denote the 5-tuple as a *relaxed abduction system* and define the *relaxed abduction problem* of such a system to be the task of determining all pairs (A, O) that solve it. However, as we only consider this one problem and want to stay aligned with existing literature in this field, we use the term *relaxed abduction problem* for the 5-tuple as well.

[3]For diagnostics, a pre-solution is therefore a pair of diagnoses and symptoms such that the diagnoses are consistent and entail the symptoms under the domain model.

$A \lessdot A'$. The preorder $\preceq_{\mathcal{O}}$ is called dually irredundant *if and only if for all pre-solutions* $(A, O), (A', O')$ *to* **RAP**: *If* $O \supsetneq O'$, *then* $O \prec_{\mathcal{O}} O'$.

These stricter criteria on the admissible preorders naturally give rise to a special class of relaxed abduction problems:

Definition 3.3 (Strict relaxed abduction problem)

A strict relaxed abduction problem (or strict **RAP***) is a relaxed abduction problem* **RAP** $= (\mathcal{T}, \mathcal{A}, \mathcal{O}, \preceq_{\mathcal{A}}, \preceq_{\mathcal{O}})$*, where* $\preceq_{\mathcal{A}}$ *is irredundant and* $\preceq_{\mathcal{O}}$ *is dually irredundant.*

In spite of the general nature, already the basic (anti-) monotonicity requirements imposed on the preference relations in Definition 3.1 suffice to establish two central properties of relaxed abduction, namely conservativeness and weak optimality. The fact that relaxed abduction conservatively extends standard logic-based abduction guarantees that, under natural conditions, our extended approach reproduces all solutions to the corresponding standard abduction problem. Intuitively, this means that no "perfect" solution explaining all observations is lost when applying the relaxation.

Proposition 3.1 (Conservativeness of the relaxation)

Let **RAP** $= (\mathcal{T}, \mathcal{A}, \mathcal{O}, \preceq_{\mathcal{A}}, \preceq_{\mathcal{O}})$ *be a relaxed abduction problem, with the corresponding preferential abduction problem* **PrefAP** $= (\mathcal{T}, \mathcal{A}, \mathcal{O}, \preceq_{\mathcal{A}})$*. Then, the following properties hold:*

a) $A \models$ **PrefAP** *if and only if* $(A, \mathcal{O}) \models$ **RAP**

b) $|Sol_{\textbf{PrefAP}}| \leq |Sol_{\textbf{RAP}}|$*, and* $|Sol_{\textbf{PrefAP}}| < |Sol_{\textbf{RAP}}|$ *if and only if* $\mathcal{T} \not\models \mathcal{O}$

Proof. For the proof the "only if" part of property a) assume $A \models$ **PrefAP**. Then, by Definition 2.3, $\mathcal{T} \cup A$ is consistent, $\mathcal{T} \cup A \models \mathcal{O}$, and A is $\preceq$-minimal among all such pre-solutions. Since $\preceq_{\mathcal{O}}$ is anti-monotone for set inclusion by requirement (i. e. prefers supersets over subsets), $\mathcal{O}$ is naturally the $\preceq_{\mathcal{O}}$-minimal element of $\mathcal{P}(\mathcal{O})$. $(A, \mathcal{O})$ is therefore $\preceq$-minimal, yielding $(A, \mathcal{O}) \models$ **RAP**.

For the reverse direction of a), if $(A, \mathcal{O}) \models$ **RAP** then, by Definition 3.1, $\mathcal{T} \cup A$ is consistent, $\mathcal{T} \cup A \models \mathcal{O}$, and $(A, \mathcal{O})$ is $\preceq$-minimal in the set of these pre-solutions. It follows that A must be a pre-solution for **PrefAP** and it remains to show that A is $\preceq_{\mathcal{A}}$-minimal for the set of pre-solutions. Assume on the contrary $\exists A' \subseteq \mathcal{A}$ such that A' is also a pre-solution, and $A' \prec_{\mathcal{A}} A$. By definition of $\preceq$ it follows that $(A', \mathcal{O}) \prec (A, \mathcal{O})$, which is a contradiction to the premise. A must therefore be a $\preceq_{\mathcal{A}}$-minimal element of $\mathcal{A}$ w. r. t. $\preceq_{\mathcal{A}}$.

We now prove property b): $|Sol_{\mathbf{PrefAP}}| \leq |Sol_{\mathbf{RAP}}|$ follows directly from a). To show that the strict relation holds if any only if $\mathcal{T} \not\models \mathcal{O}$, note that there is always a (possibly empty) set of observations which are trivially entailed by $\mathcal{T}$, i.e. $Sol_{\mathbf{RAP}}$ always contains a "trivial" solution $(\emptyset, O)$ for $\emptyset \subseteq O \subseteq \mathcal{O}$. If $\mathcal{T} \not\models \mathcal{O}$, then $O \subsetneq \mathcal{O}$ and therefore $\emptyset \not\models \mathbf{PrefAP}$. $Sol_{\mathbf{RAP}}$ then contains at least one element that has no correspondence in $Sol_{\mathbf{PrefAP}}$, justifying the "if" part of the second claim. For the "only if" part, assume $\mathcal{T} \models \mathcal{O}$. Then $|Sol_{\mathbf{PrefAP}}| = |\{\emptyset\}| = 1 = |\{(\emptyset, \mathcal{O})\}| = |Sol_{\mathbf{RAP}}|$. Taking the contraposition justifies the claim. $\qquad\square$

Due to the definition of $\preceq$ as Cartesian product of the preorders $\preceq_{\mathcal{A}}$ and $\preceq_{\mathcal{O}}$, it can be shown that all solutions to **RAP** are (weakly) Pareto-optimal. This observation guarantees that each of the additional solutions a **RAP** offers over the corresponding **PrefAP** is indeed valuable as it represents an optimal tradeoff between simplicity and consilience.

Proposition 3.2 (Weak Pareto-optimality of RAP-solutions)
Let $\mathbf{RAP} = (\mathcal{T}, \mathcal{A}, \mathcal{O}, \preceq_{\mathcal{A}}, \preceq_{\mathcal{O}})$ *be a relaxed abduction problem. It then holds that* $(A^*, O^*) \models \mathbf{RAP}$ *if and only if* (A^*, O^*) *is a weakly Pareto-optimal element of the set of pre-solutions* $\{(A, O) \in \mathcal{P}(\mathcal{A}) \times \mathcal{P}(\mathcal{O}) \mid \mathcal{T} \cup A \models O \wedge \mathcal{T} \cup A \not\models \bot\}$ *subject to* $\preceq_{\mathcal{A}}$ *and* $\preceq_{\mathcal{O}}$.

Proof. For the proof of the "only if" direction, assume $(A^*, O^*) \models \mathbf{RAP}$. Then, by Definition 3.1, (A^*, O^*) is a pre-solution, and (A^*, O^*) is a minimal element in the set of pre-solutions. Assume that (A^*, O^*) is not Pareto-optimal, and let (A', O') be another pre-solution that dominates (A^*, O^*). W.l.o.g we assume that $A' \prec_{\mathcal{A}} A^*$ and $O' \preceq_{\mathcal{O}} O^*$ (the argument for $A' \preceq_{\mathcal{A}} A^*$ and $O' \prec_{\mathcal{O}} O^*$ is structurally similar). Then $(A', O') \prec (A^*, O^*)$ according to Definition 3.1, which contradicts $\preceq$-minimality of (A^*, O^*). (A^*, O^*) must therefore be a Pareto-optimal element of the set of pre-solutions.
For the reverse, let (A^*, O^*) be a Pareto-optimal pre-solution, it remains to show that (A^*, O^*) is $\preceq$-minimal. Assume $(A', O') \models \mathbf{RAP}$ such that $(A', O') \prec (A^*, O^*)$, it then follows w.l.o.g. that $A' \prec_{\mathcal{A}} A^*$ and $O' \preceq_{\mathcal{O}} O^*$ which is a contradiction to the Pareto-optimality of (A^*, O^*) (again, the argument for the other choice is structurally similar). Conclusively, (A', O') must be $\preceq$-minimal. $\qquad\square$

As we show next, if **RAP** is strict, the maximum cardinality of $Sol_{\mathbf{RAP}}$ shrinks significantly, as for each set of assumptions there can be at most one solution.

Proposition 3.3 (*Sol*$_{\mathbf{RAP}}$ **size limit for strict RAP**)
Let $\mathbf{RAP} = (\mathcal{T}, \mathcal{A}, \mathcal{O}, \preceq_A, \preceq_O)$ *be a strict relaxed abduction problem. Then:
For all* $A \subseteq \mathcal{A}$: $|\{O \subseteq \mathcal{O} \mid (A, O) \in Sol_{\mathbf{RAP}}\}| \leq 1$.

Proof. Let $A \in \mathcal{P}(\mathcal{A})$, and assume there are $O_1, O_2 \in \mathcal{P}(\mathcal{O})$ such that $(A, O_1) \models \mathbf{RAP}$, $(A, O_2) \models \mathbf{RAP}$, and $O_1 \neq O_2$. Then $(A, O_1), (A, O_2)$ are both $\preceq$-minimal, $\mathcal{T} \cup A \not\models \bot$, $\mathcal{T} \cup A \models O_1$ and $\mathcal{T} \cup A \models O_2$. From the latter it follows that $\mathcal{T} \cup A \models O_1 \cup O_2$ but $O_1 \subsetneq O_1 \cup O_2$ since $O_1 \neq O_2$. Therefore, by definition of dual irredundancy, we get $O_1 \cup O_2 \preceq_O O_1$, contradicting the minimality of (A, O_1). $\qquad\square$

Summing up, already the general framework of relaxed abduction laid out so far fulfils the framework-oriented requirements formulated in Section 1.2: It is based on logics without being restricted to propositional ones, which allows to state relational information (Requirement **R1**). While missing observations are handled by the use of abduction, modelling flaws are handled by ignoring observations which cannot be explained based on the current domain ontology, therefore meeting Requirements **R2** and **R3**. Moreover, even if there are solutions that completely explain $\mathcal{O}$, relaxed abduction realizes a tradeoff between simplicity and consilience as demanded by Requirement **R5** (even more, all such tradeoffs are realized). Finally, Requirement **R4** of not producing irrelevant solutions is met by dropping any candidates which are not Pareto-optimal.

Moreover, logic-based abduction (Definitions 2.2 and 2.3) can be reduced straightforwardly to relaxed abduction by determining all relaxed solutions, and then dropping all pairs (A, O) where $O \neq \mathcal{O}$.[4] Complexity results on logic-based abduction therefore carry over to our relaxation as lower bounds. Eiter & Gottlob (1995) and Bienvenu (2008) provide a detailed overview of the plenty of results regarding the complexity of logic-based abduction currently known, showing that it crucially depends both on the choice of logic as well as on the preference criterion employed (if, for instance, the language permits the use of variables, additional effort for unification and instantiation may occur). As a general guideline, except from very simple logics, abduction is typically (significantly) harder than deduction; employing ordering criteria over the solutions generally leads to an increase in complexity as well. Moreover, determining whether a certain assumption is relevant for a solution (the so-called *support selection task*) has been identified as a hard problem lying at the core of all forms of abductive and

---

[4]Note however that this reduction is *not* polynomial as we may have to investigate an exponential number of pairs (A, O).

default reasoning in Selman & Levesque (1990). This makes abduction an NP-hard problem even for propositional Horn theories. Putting the pieces together, PTIME algorithms for the problem of relaxed abduction are not to be expected.

Before providing a more detailed complexity analysis in Section 3.2.4, we return to the contents of Section 3.1 and show how diagnostics can be represented and solved in the framework of relaxed abduction.

3.2.3 Instantiating Relaxed Abduction: Mapping to Diagnostics

Based on the $\mathcal{EL}^+$ formalisation of diagnostics introduced in Section 3.1.2, flexible diagnostic reasoning can be realised by means of relaxed abduction for both causal and anti-causal representation of system knowledge.

Causal Diagnostic Knowledge

Causal diagnostic knowledge is typically available in situations where the internal processes of the system under consideration are well known. We therefore assume given in addition to $\mathcal{M}_{sys}$ and $\mathcal{M}_{cust}$ an ontology $\mathcal{M}_{caus}$ expressing how fault modes manifest themselves as symptoms of the various dependent components. Moreover, we denote by $\mathcal{M}_{obs}$ those axioms in $\mathcal{M}_{sym}$ that reflect the set of symptoms currently being observed, i.e. $\mathcal{M}_{obs} \subseteq \mathcal{M}_{sym}$.

To solve this diagnostic task, we instantiate the relaxed abduction problem $\mathbf{RAP_{caus}} = (\mathcal{T}, \mathcal{A}, \mathcal{O}, \preceq_{\mathcal{A}}, \preceq_{\mathcal{O}})$ with $\mathcal{T} = \mathcal{M}_{sys} \cup \mathcal{M}_{cust} \cup \mathcal{M}_{caus}$, $\mathcal{O} = \mathcal{M}_{obs}$, $\mathcal{A} = \mathcal{M}_{diag}$, $\preceq_{\mathcal{A}} = \subseteq$, and $\preceq_{\mathcal{O}} = \supseteq$. Solving $\mathbf{RAP_{caus}}$ with either of the algorithms introduced before yields $Sol_{\mathbf{RAP_{caus}}} = \{(A_1, O_1), \ldots, (A_n, O_n)\}$. Axioms in A_i and in O_i have the respective forms $\langle \mathsf{Sys_i} \rangle \sqsubseteq \exists\, \mathsf{operatesIn}\,.$ $\langle \mathsf{Fault} \rangle$ and $\langle \mathsf{Sys_i} \rangle \sqsubseteq \exists\, \mathsf{shows}\,.\, \langle \mathsf{Sym_{i,j}} \rangle$, indicating which faulty system behaviour has to be assumed (A_i) to explain a certain subset of the observed symptoms (O_i) based on the causal knowledge base. The preorders $\preceq_{\mathcal{A}}$ and $\preceq_{\mathcal{O}}$ are exemplarily instantiated with subset relationships to encourage dropping superfluous assumptions and adding observations; however entailment-based preferences could be used straightforwardly as well. This way, we employ the relaxed abduction mechanism to obtain optimal fault explanations for possibly incomplete sets of symptomatic observations on systems, even in situations where not all of the symptomatic observations can be explained at once or simpler explanations can be found by omitting certain observations.

Anti-Causal Diagnostic Knowledge

If an anti-causal diagnostic knowledge base $\mathcal{M}_{\overline{caus}}$ is given, the natural way of performing diagnostics is to apply deductive reasoning over the system knowledge. This directly derives the faults responsible for the observed symptoms according to the direction of the implication from symptoms to fault. However, deriving faults by means of deduction requires complete observational data, this prerequisite is typically not met in practice due to defunct sensors, for instance. In such a situation it is desirable to also obtain hypothetical diagnoses which indicate a potential fault that would be derivable from the system knowledge if only a (preferably small) number of additional symptomatic observations was made. Again, we denote by $\mathcal{M}_{obs}$ those axioms in $\mathcal{M}_{sym}$ that reflect the currently observed exhibition of symptoms, $\mathcal{M}_{obs} \subseteq \mathcal{M}_{sym}$.

Relaxed abduction can be instantiated to solve this task as well by letting $\mathcal{T} = \mathcal{M}_{sys} \cup \mathcal{M}_{cust} \cup \mathcal{M}_{\overline{caus}} \cup \mathcal{M}_{obs}$, $\mathcal{O} = \mathcal{M}_{diag}$, $\mathcal{A} = \mathcal{M}_{sym}$, $\preceq_{\mathcal{A}} = \subseteq$, and $\preceq_{\mathcal{O}} = \supseteq$. That is, we add to the background knowledge in $\mathcal{T}$ the actual symptomatic observations as given statements, let $\mathcal{O}$ be the possible diagnoses to be explained, and let $\mathcal{A}$ be the set of all potential symptomatic observations. The solution to the resulting relaxed abduction problem $\mathbf{RAP}_{\overline{caus}} = (\mathcal{T}, \mathcal{A}, \mathcal{O}, \preceq_{\mathcal{A}}, \preceq_{\mathcal{O}})$ is then given by $Sol_{\mathbf{RAP}_{\overline{caus}}} = \{(A_1, O_1), \ldots, (A_n, O_n)\}$. Now, axioms in A_i and in O_i have the form $\langle \mathsf{Sys_i} \rangle \sqsubseteq \exists \mathsf{shows}\,.\,\langle \mathsf{Sym_{i,j}} \rangle$ and $\langle \mathsf{Sys_i} \rangle \sqsubseteq \exists \mathsf{operatesIn}\,.\,\langle \mathsf{Fault} \rangle$ respectively, indicating which additional symptomatic observations have to be assumed (A_i) to explain the possible (or hypothetical) diagnoses (O_i), given the anti-causal knowledge. By this, we obtain hypothetical diagnoses that explain potential system faults with (parts of) the symptomatic observations combined with additionally assumed symptoms. The ordering mechanism of relaxed abduction provides for a ranking of such diagnoses, preferring those that require fewest additional assumptions and explain most of the observations (again, other criteria such as the entailment-based one could be used as well).

Table 3.1 summarises these results. Here, diagnosis denotes the process of diagnostic inference as outlined for both representation options. By validation or prognosis, we understand the derivation of additional expected observations from a set of diagnoses. This can be useful on the one hand to guide technicians in searching for additional symptoms to verify or falsify a diagnosis, but also for predicting which symptoms are expected to happen in the future, given the inferred diagnosis.

Table 3.1: Relation between type of knowledge modelling, reasoning method, and application

Modelling	Reasoning method	Typical application
Causal	Deduction	Validation/Prognosis
	Abduction	Diagnosis
	Relaxed Abduction	Validation/Prognosis & Diagnosis
Anti-causal	Deduction	Diagnosis
	Abduction	Validation/Prognosis
	Relaxed Abduction	Validation/Prognosis & Diagnosis

3.2.4 Instantiating Relaxed Abduction: Choices for $\preceq_{\mathcal{A}}$ and $\preceq_{\mathcal{O}}$

Until now we have focused on the general formulation of relaxed abduction without considering the effects of concrete choices for the preorders $\preceq_{\mathcal{A}}$ and $\preceq_{\mathcal{O}}$. This section investigates a number of natural choices. Note that while we only consider uniform combinations where $\preceq_{\mathcal{A}}$ and $\preceq_{\mathcal{O}}$ are of the same family (e. g. both based on set inclusion), this is not enforced by the definition of relaxed abduction. Based on domain requirements, members from different families may be chosen, their interaction and properties however should be considered carefully in advance.

Inclusion-based Orders

The definition of relaxed abduction requires $\preceq_{\mathcal{A}}$ to be monotone and $\preceq_{\mathcal{O}}$ to be anti-monotone for set inclusion. Therefore, the simplest and most straightforward instantiation of the framework is given by $\mathbf{RAP} = (\mathcal{T}, \mathcal{A}, \mathcal{O}, \subseteq, \supseteq)$, denoted by $\mathbf{sRAP}$. Here simplicity is identified with subset-minimality of the assumption set, which can be understood as removing elements from A one by one as long as no observations are lost. Consilience is identified accordingly with superset-maximality of the observation set, intuitively corresponding to the stepwise addition of observations which can be explained by the current set of assumptions. Note that both $\subseteq$ and $\supseteq$ fulfil the antisymmetry property and are hence partial orders.

As $\preceq$-minimality is a necessary requirement for any solution, the size of $Sol_{\mathbf{RAP}}$ is dependent on the concrete choice of $\preceq_{\mathcal{A}}$ and $\preceq_{\mathcal{O}}$, respectively. If both preorders were chosen to be $\emptyset$, i. e. if no dominance were employed at all, this would obviously allow for up to $2^{|\mathcal{A}|+|\mathcal{O}|}$ solutions (the effective number depending on the axiom sets $\mathcal{T}$, $\mathcal{A}$, and $\mathcal{O}$). The use of inclusion-based orders already reduces this worst case estimate significantly:

Proposition 3.4 (Properties for inclusion-based orders)
Let $\textbf{RAP} = (\mathcal{T}, \mathcal{A}, \mathcal{O}, \subseteq, \supseteq)$ *be an inclusion-based relaxed abduction problem (or* $\textbf{sRAP}$*). The following properties hold:*

    a) $\forall A \subseteq \mathcal{A} : |\{O \subseteq \mathcal{O} \mid (A, O) \in Sol_{\textbf{RAP}}\}| \leq 1$

    b) $\forall O \subseteq \mathcal{O} : |\{A \subseteq \mathcal{A} \mid (A, O) \in Sol_{\textbf{RAP}}\}| \leq \binom{|\mathcal{A}|}{\lfloor |\mathcal{A}|/2 \rfloor} \approx 2^{|\mathcal{A}|}/\sqrt{\frac{\pi}{2} \cdot |\mathcal{A}|}$

    c) $|Sol_{\textbf{RAP}}| \leq \min\left\{2^{|\mathcal{A}|}, \binom{|\mathcal{A}|}{\lfloor |\mathcal{A}|/2 \rfloor} \cdot 2^{|\mathcal{O}|}\right\} \in O\left(2^{|\mathcal{A}|} \cdot \min\left\{\frac{2^{|\mathcal{O}|}}{\sqrt{|\mathcal{A}|}}, 1\right\}\right)$

    d) $\exists\, \mathcal{T}, \mathcal{A}, \mathcal{O} : |Sol_{\textbf{RAP}}| = 2^{|\mathcal{A}|}$

Proof. a) directly follows from Proposition 3.3 and the fact that $\preceq_A$ ($\preceq_O$) is indeed irredundant (dually irredundant) by definition (i.e. any inclusion-based **RAP** is strict). To show b) consider $O^* \subseteq \mathcal{O}$ fixed. To prove the upper bound, note that all sets (A, O) have to be solutions and O is fixed (due to the quantifier). The problem is therefore to determine the maximum size of a so-called antichain over $\mathcal{A}$ of sets A_i, where no A_i may be a superset of another A_j (as otherwise dominance would occur. According to Sperner's theorem (Sperner, 1928), this maximum size is given by $\binom{|\mathcal{A}|}{\lfloor |\mathcal{A}|/2 \rfloor}$. Next, we derive an upper bound on the value of this formula. Setting $m := \lfloor \frac{|\mathcal{A}|}{2} \rfloor$, it can be shown that $\binom{2m}{m} \leq \binom{|\mathcal{A}|}{\lfloor |\mathcal{A}|/2 \rfloor} < 2 \cdot \binom{2m}{m}$: If $|\mathcal{A}|$ is even, then $|\mathcal{A}| = 2m$ and therefore $\binom{2m}{m} = \binom{|\mathcal{A}|}{\lfloor |\mathcal{A}|/2 \rfloor}$. If on the other hand $|\mathcal{A}|$ is odd, then $|\mathcal{A}| = 2m + 1$, which implies $\binom{|\mathcal{A}|}{\lfloor |\mathcal{A}|/2 \rfloor} = \binom{2m+1}{m}$. By simple maths $\binom{2m+1}{m} = \binom{2m+1}{m+1} = \binom{2m}{m} + \binom{2m}{m+1} < 2 \cdot \binom{2m}{m}$. Taken together, we get that the maximum size of such an antichain is bounded form above by $2 \cdot \binom{2m}{m} = 2 \cdot \frac{(2m)!}{(m!)^2}$. Approximating the factorial $m!$ by the first two summands of Stirling's series, i.e. $m! \approx \sqrt{\pi/3 \cdot (6m + 1)} \left(\frac{m}{e}\right)^m$, yields the estimate $\binom{|\mathcal{A}|}{\lfloor |\mathcal{A}|/2 \rfloor} \approx 2^{|\mathcal{A}|+1}/\sqrt{\frac{\pi}{2} \cdot |\mathcal{A}|}$ with an error of less than 0.1% for $|\mathcal{A}| > 4$ (see Fellner, 1968, §2.9). This concludes the proof.

Property c) directly follows from properties a) and b), and the fact that there are $2^{|\mathcal{A}|}$ ($2^{|\mathcal{O}|}$) distinct subsets of $\mathcal{A}$ ($\mathcal{O}$).

To prove d), take $\textbf{RAP} = (\emptyset, \mathcal{A}, \mathcal{A}, \subseteq, \supseteq)$ and let $\mathcal{A}$ be logically independent (see Definition 2.1). Then each observation can be explained in exactly one way, namely by assuming it. This results in $|Sol_{\textbf{RAP}}| = |\{(A, A) \mid A \subseteq \mathcal{A}\}| = 2^{|\mathcal{A}|}$. This shows the tightness of the bound for the typical case $\pi/2 \cdot |\mathcal{A}| \leq 2^{2 \cdot |\mathcal{O}|}$. $\qquad\square$

The key result from Proposition 3.4 is that for solving relaxed abduction problems, the worst-case upper bound on runtime is exponential runtime in the number of abducibles (and in the number of observations, yet dampened by a factor of $\sqrt{|\mathcal{A}|}$). It is not hard to show (and in accordance with well-known results on goal-directed reasoning in general) that there exist cases that realize this worst case scenario. Solving a relaxed abduction problem is therefore a computationally non-trivial task. Nevertheless, the following example shows that already the natural, subsumption-based choice of (pre)orders leads to a significant reduction of the solution set size (and, therefore, computation time). Moreover, the Proposition 3.4 demonstrates that this instantiation already provides a solution to the problem posed in Example 2.1:

Example 3.1 (Diagnostics over incomplete domain formalisations (cont.))
As before, we consider the domain ontology $\mathcal{T}$ given by

$$\mathsf{MCU} \sqsubseteq \exists \, \mathsf{partOf} \, . \, \mathsf{System}$$
$$\mathsf{Communications} \sqsubseteq \exists \, \mathsf{subsystemOf} \, . \, \mathsf{System}$$
$$\mathsf{Transportation} \sqsubseteq \exists \, \mathsf{subsystemOf} \, . \, \mathsf{System}$$
$$\mathsf{PROFINET} \sqsubseteq \exists \, \mathsf{belongsTo} \, . \, \mathsf{Communications}$$
$$\mathsf{Gripper} \sqsubseteq \exists \, \mathsf{belongsTo} \, . \, \mathsf{Transportation}$$
$$\mathsf{Conveyor} \sqsubseteq \exists \, \mathsf{belongsTo} \, . \, \mathsf{Transportation}$$
$$\mathsf{belongsTo} \circ \mathsf{subsystemOf} \sqsubseteq \mathsf{partOf}$$
$$\mathsf{MCU} \sqcap \exists \, \mathsf{partOf} \, . \, (\exists \, \mathsf{operatesIn} \, . \, \mathsf{PowerSupplyFluctuations})$$
$$\sqsubseteq \exists \, \mathsf{shows} \, . \, \mathsf{IntermittentOutages}$$
$$\mathsf{PROFINET} \sqcap \exists \, \mathsf{partOf} \, . \, (\exists \, \mathsf{operatesIn} \, . \, \mathsf{PowerSupplyFluctuations})$$
$$\sqsubseteq \exists \, \mathsf{shows} \, . \, \mathsf{SendingReceivingOK}$$
$$\mathsf{Gripper} \sqcap \exists \, \mathsf{partOf} \, . \, (\exists \, \mathsf{operatesIn} \, . \, \mathsf{PowerSupplyFluctuations})$$
$$\sqsubseteq \exists \, \mathsf{shows} \, . \, \mathsf{FullyFunctional}$$
$$\mathsf{MCU} \sqcap \exists \, \mathsf{partOf} \, . \, (\exists \, \mathsf{operatesIn} \, . \, \mathsf{ControlSWMalfunction})$$
$$\sqsubseteq \exists \, \mathsf{shows} \, . \, \mathsf{IntermittentOutages}$$
$$\mathsf{Conveyor} \sqcap \exists \, \mathsf{partOf} \, . \, (\exists \, \mathsf{operatesIn} \, . \, \mathsf{ControlSWMalfunction})$$
$$\sqsubseteq \exists \, \mathsf{shows} \, . \, \mathsf{IrregularMovements}$$
$$\mathsf{Gripper} \sqcap \exists \, \mathsf{partOf} \, . \, (\exists \, \mathsf{operatesIn} \, . \, \mathsf{ControlSWMalfunction})$$
$$\sqsubseteq \exists \, \mathsf{shows} \, . \, \mathsf{IrregularMovements}$$

Note that, with respect to the ontology hierarchy outlined in Sections 3.1.2 and 3.2.3, the upper seven axioms would form $\mathcal{M}_{cust}$ whereas the lower six axioms make up the ontology $\mathcal{M}_{caus}$. As in the previous example, the sets of abducibles and observations (including the new vibration symptom) are given by:

$$\mathcal{A}: \quad \text{System} \sqsubseteq \exists\, \text{operatesIn}\,.\, \text{PowerSupplyFluctuations}$$
$$\text{System} \sqsubseteq \exists\, \text{operatesIn}\,.\, \text{ControlSWMalfunction}$$
$$\mathcal{O}': \quad \text{MCU} \sqsubseteq \exists\, \text{shows}\,.\, \text{IntermittentOutages}$$
$$\text{Gripper} \sqsubseteq \exists\, \text{shows}\,.\, \text{FullyFunctional}$$
$$\text{MCU} \sqsubseteq \exists\, \text{shows}\,.\, \text{LowFreqVibrations}$$

Here the set of abducibles corresponds to $\mathcal{M}_{diag}$, and the set of observations is $\mathcal{M}_{obs}$, a subset of the (unspecified) set $\mathcal{M}_{sym}$ of possible symptom statements.

Using set inclusion as ordering criterion, we define $A \preceq_A A'$ if and only if $A \subseteq A'$ and $O \preceq_\mathcal{O} O'$ if and only if $O \supseteq O'$. As intended, the resulting order $\preceq$ gives rise to a solution (A_1, O_1) defined by:

$$A_1 = \{\text{System} \sqsubseteq \exists\, \text{operatesIn}\,.\, \text{PowerSupplyFluctuations}\}$$
$$O_1 = \{\text{MCU} \sqsubseteq \exists\, \text{shows}\,.\, \text{IntermittentOutages},$$
$$\text{Gripper} \sqsubseteq \exists\, \text{shows}\,.\, \text{FullyFunctional}\}$$

This solution explains all observations but the vibrations and only requires to assume the diagnosis, namely a fluctuating power supply. However, there's additional solution candidates to be explored taking the second diagnosis candidate defined in $\mathcal{M}_{diag}$ into account, namely (A_2, O_2)

$$A_2 = \{\text{System} \sqsubseteq \exists\, \text{operatesIn}\,.\, \text{ControlSWMalfunction}\}$$
$$O_2 = \{\text{MCU} \sqsubseteq \exists\, \text{shows}\,.\, \text{IntermittentOutages}\}$$

and (A_3, O_3):

$$A_3 = \{\text{System} \sqsubseteq \exists\, \text{operatesIn}\,.\, \text{PowerSupplyFluctuations},$$
$$\text{System} \sqsubseteq \exists\, \text{operatesIn}\,.\, \text{ControlSWMalfunction}\}$$
$$O_3 = \{\text{MCU} \sqsubseteq \exists\, \text{shows}\,.\, \text{IntermittentOutages},$$
$$\text{Gripper} \sqsubseteq \exists\, \text{shows}\,.\, \text{FullyFunctional}\}$$

The latter candidate is however not a valid solution since it is dominated by the first one, $(A_1, O_1) \prec (A_3, O_3)$. Based on the information given, the only other minimal solution is $(\emptyset, \emptyset)$, completing the solution set Sol.

One of the favourable features of this definition of relaxed abduction is the simplistic definition. However, although inclusion-based orders fulfil the property of irredundancy, they exhibit a largely syntax-dependent behaviour. To address this potential drawback, we next take a look at a class of relaxed abduction problems that rely on semantic entailment.

Entailment-based Orders

Preference relations based on semantic entailment are a natural choice in the context of a logic-based reasoning framework such as relaxed abduction. As entailment is naturally an anti-monotonic preorder for set inclusion ($S \subseteq S'$ implies $S \models\mid S'$), it is tempting to define simple entailment-based relaxed abduction by $\mathbf{seRAP} = (\mathcal{T}, \mathcal{A}, \mathcal{O}, \models\mid, \models)$. However, it can be shown that in the presence of assertional axioms, results from $\mathbf{seRAP}$ are incomparable to $\mathbf{sRAP}$ results:

Example 3.2 (Subset- and simple entailment-based RAP are incomparable)
Let $\mathcal{T} = \emptyset, \mathcal{A} = \{\mathsf{A}(a), \mathsf{A} \sqcap \mathsf{A}(a)\}, \mathcal{O} = \{\mathsf{A}(a)\}$. Then $(\mathcal{A}, \mathcal{O})$ is a solution to the corresponding seRAP, but not to sRAP.
Conversely, for $\mathcal{A}' = \{\mathsf{A} \sqcup \mathsf{B}(a), \mathsf{A}(a)\}, \mathcal{O} = \{\mathsf{A} \sqcup \mathsf{B}(a)\}$, we have $(\{\mathsf{A}(a)\}, \mathcal{O})$ as a solution to sRAP, but not to seRAP.

Therefore, to realize our goal of reducing syntax sensitivity of the inclusion-based approach along with the size of the solution set, we define a slightly more complex preference relation for entailment-based orders:

Definition 3.4 (Entailment-based relaxed abduction problem)
*An entailment-based **RAP**, denoted **eRAP**, is a relaxed abduction problem with $\preceq_\mathcal{A}$, $\preceq_\mathcal{O}$ defined as follows:*

- *$A_1 \preceq_\mathcal{A} A_2$ if and only if*
  $(A_1 \equiv A_2$ and $A_1 \subseteq A_2)$ or $(A_1 \not\equiv A_2$ and $A_1 \models\mid A_2)$
- *$O_1 \preceq_\mathcal{O} O_2$ if and only if*
  $(O_1 \equiv O_2$ and $O_1 \supseteq O_2)$ or $(O_1 \not\equiv O_2$ and $O_1 \models O_2)$

The idea of this definition is to partition the sets of hypotheses and the observation sets into equivalent classes and then order sets within the classes w. r. t. to set inclusion. The entailment relation makes up the outer structure for the preference whereas the inclusion ordering gives the inner structure for ordering the classes. Hence the approach we follow in the definition of **eRAP** is comparable to the prioritization approach proposed by (Eiter &

Gottlob, 1995, p.5). As we show next, $\preceq_A$ and $\preceq_O$ are indeed admissible preorders.

Lemma 3.5 ((Anti-)monotonicity of prcorders)
$\preceq_A$ and$\preceq_O$ are preorders that are monotone (anti-monotone) for set inclusion.

Proof. $\preceq_A$ is clearly reflexive. For showing transitivity, assume $X \preceq_A Y \preceq_A Z$. There are four combinations possible: i) $X \equiv Y, Y \equiv Z$: Then $X \equiv Z$ and $X \subseteq Y \subseteq Z$, hence $X \preceq_A Z$ due to the first disjunct of the definition. ii) $X \equiv Y, Y \not\equiv Z$, then $X \not\equiv Z$ and $Y \not\equiv Z$ and hence also $X \not\equiv Z$. So $X \preceq_A Z$ due to the second disjunct of the definition. iii) $X \not\equiv Y, Y \equiv Z$: Then $X \not\equiv Z$ and $X \not\equiv Y$ and hence also $X \not\equiv Z$, so $X \preceq_A Z$ due to the second disjunct of the definition. iv) $X \not\equiv Y, Y \not\equiv Z$, then $X \not\equiv Y$ and $Y \not\equiv Z$, hence also $X \not\equiv Z$. Now, it could not be the case that $X \equiv Z$, as that would entail that $Z \not\equiv X \not\equiv Y$ so $Z \equiv Y$, hence due to the second disjunct we again have $X \preceq_A Z$. In order to show monotonicity, let $A_1 \subseteq A_2$. If $A_1 \equiv A_2$, then $A_1 \preceq_A A_2$ follows from the first disjunct of the definition, otherwise it follows from the second.
The proof for $\preceq_O$ is dual. □

Based on the notion of a logically independent set introduced in Definition 2.1, we now derive basic properties of this instantiation:

Proposition 3.6 (Properties for entailment-based orders)
Let $\mathbf{sRAP} = (\mathcal{T}, \mathcal{A}, \mathcal{O}, \subseteq, \supseteq)$ be an inclusion-based relaxed abduction problem, and $\mathbf{eRAP} = (\mathcal{T}, \mathcal{A}, \mathcal{O}, \preceq_A, \preceq_O)$ an entailment-based relaxed abduction problem over the same axiom sets. Then, the following properties hold:

a) $Sol_{\mathbf{eRAP}} \subseteq Sol_{\mathbf{sRAP}}$

b) If $\mathcal{A}$ and $\mathcal{O}$ both are logically independent, then $Sol_{\mathbf{eRAP}} = Sol_{\mathbf{sRAP}}$.

c) $|Sol_{\mathbf{eRAP}}| \leq \min\left\{2^{|\mathcal{A}|}, \binom{|\mathcal{A}|}{\lfloor|\mathcal{A}|/2\rfloor} \cdot 2^{|\mathcal{O}|}\right\} \in O\left(2^{|\mathcal{A}|} \cdot \min\left\{\frac{2^{|\mathcal{O}|}}{\sqrt{|\mathcal{A}|}}, 1\right\}\right)$

d) There are $\mathcal{T}, \mathcal{A}, \mathcal{O}$ s.t. $|Sol_{\mathbf{eRAP}}| = 2^{|\mathcal{A}|}$

Proof. To simplify notation we denote the dominance relation induced by $\mathbf{eRAP}$ by $\preceq^e$, and by $\preceq^s$ its counterpart from $\mathbf{sRAP}$.
For the proof of property a), let $(A, O) \in Sol_{\mathbf{eRAP}}$ and assume $(A, O) \notin Sol_{\mathbf{sRAP}}$. Since the problems differ only in the dominance relations used, there must then exist $(A', O') \in Sol_{\mathbf{sRAP}}$ such that $(A', O') \prec^s (A, O)$ which is by Definition 3.1 equivalent to $A' \subset A \wedge O' \supseteq O$ or $A' \subseteq A \wedge O' \supset O$. The former case implies $A' \not\equiv A$, $O' \models O$ and $A' \neq A$, which gives $(A', O') \prec^e$

(A, O) and leads to a contradiction with the assumption. The second case is falsified similarly, concluding the proof.

To prove b) it remains to show that $Sol_{\mathbf{eRAP}} = Sol_{\mathbf{sRAP}}$ if both $\mathcal{A}$ and $\mathcal{O}$ are logically independent. From Definition 2.1 and the monotonicity of $\models$ it follows that $\forall A, A' \subseteq \mathcal{A} : A' \models A \leftrightarrow A' = A$ and $\forall O, O' \subseteq \mathcal{O} :$ $O' \models O \leftrightarrow O' = O$ are valid. Entailment and equivalence can therefore be used interchangeably for logically independent sets $\mathcal{A}$ and $\mathcal{O}$, giving $Sol_{\mathbf{eRAP}} = Sol_{\mathbf{sRAP}}$.[5]

Property c) follows directly from a) in combination with Proposition 3.4. The proof of property d) is identical to the one for Proposition 3.4 d). $\square$

Note that the concrete benefit of using the **eRAP** ordering criterion depends on the structure of $\mathcal{A}$ and $\mathcal{O}$. If there is a hierarchy of assumptions getting more and more specific, the benefit may in fact be much larger than the worst-case estimate suggests. The approach can be further extended by employing entailment w. r. t. the axiom set $\mathcal{T}$ ($\models_{\mathcal{T}}$) for dominance instead of basic entailment:

Definition 3.5 ($\mathcal{T}$-entailment-based relaxed abduction problem)
An $\mathcal{T}$-entailment-based **RAP**, *denoted* **eRAP***, *is a relaxed abduction problem with* $\preceq_{\mathcal{A}}$, $\preceq_{\mathcal{O}}$ *defined as follows:*

- $A_1 \preceq_{\mathcal{A}} A_2$ *if and only if*
  $(A_1 \equiv_{\mathcal{T}} A_2$ *and* $A_1 \subseteq A_2)$ *or* $(A_1 \not\equiv_{\mathcal{T}} A_2$ *and* $A_1 \models_{\mathcal{T}} A_2)$
- $O_1 \preceq_{\mathcal{O}} O_2$ *if and only if*
  $(O_1 \equiv_{\mathcal{T}} O_2$ *and* $O_1 \supseteq O_2)$ *or* $(O_1 \not\equiv_{\mathcal{T}} O_2$ *and* $O_1 \models_{\mathcal{T}} O_2)$

The following proposition shows that this modification can be understood as a strengthening:

Proposition 3.7 ($\mathcal{T}$-strengthening for entailment-based orders)
Let **eRAP** *be an entailment-based* **RAP**, *and* **eRAP*** *a $\mathcal{T}$-entailment-based* **RAP** *over the same axiom sets. Then* $Sol_{\mathbf{eRAP}^*} \subseteq Sol_{\mathbf{eRAP}}$.

Proof. The transition from $\models$ to $\models_{\mathcal{T}}$ corresponds to a left weakening of the entailments (for the definition and correctness of the left-weakening step see Buss, 1998). Therefore if **eRAP*** $\models (A, O)$, then $\preceq_{\mathcal{T}}$-minimality of (A, O) implies $\preceq$-minimality of (A, O), justifying the claim **eRAP** $\models (A, O)$. $\square$

Thus, by employing the implicit information encoded in the axiom sets $\mathcal{T}$, $\mathcal{A}$ and $\mathcal{O}$, there is potential for a (possibly significant) reduction in the

---

[5]Note that the property of irredundancy also collapses to monotonicity under the assumption of independent sets.

number of solutions determined. Whether this reduction also results in an improvement of the total runtime crucially depends on how efficiently entailment can be tested, since there is typically a large number of dominance tests required for determining $Sol_{\mathbf{RAP}}$. Moreover, the amount of implicit information accessible in this way (and therefore the potential for optimisation) is a function of the concrete knowledge bases used but not a property of the relaxed abduction algorithm. We therefore refrain from evaluating runtime effects of this choice.

Cardinality-based Orders

Intuitively, a more significant reduction in the number of solutions is to be expected by defining dominance based on the number of elements in A and O. The corresponding instantiation of the framework is $\mathbf{RAP} = (\mathcal{T}, \mathcal{A}, \mathcal{O}, \leq^{|\cdot|}, \geq^{|\cdot|})$, with $A \leq^{|\cdot|} A'$ if and only if $|A| \leq |A'|$, and $O \geq^{|\cdot|} O'$ if and only if $|O| \geq |O'|$. Surprisingly, the maximum number of solutions is identical to the set inclusion case, as the following proposition shows.

Proposition 3.8 (Properties for cardinality-based orders)
Let $\mathbf{cRAP} = (\mathcal{T}, \mathcal{A}, \mathcal{O}, \leq^{|\cdot|}, \geq^{|\cdot|})$ and $\mathbf{sRAP} = (\mathcal{T}, \mathcal{A}, \mathcal{O}, \subseteq, \supseteq)$ be two relaxed abduction problems over the same axiom sets. The following properties hold:

a) $Sol_{\mathbf{cRAP}} \subseteq Sol_{\mathbf{sRAP}}$

b) $|Sol_{\mathbf{cRAP}}| \leq \min\left\{2^{|A|}, \binom{|A|}{\lfloor|A|/2\rfloor} \cdot 2^{|O|}\right\} \in O\left(2^{|A|} \cdot \min\left\{\frac{2^{|O|}}{\sqrt{|A|}}, 1\right\}\right)$

c) $\exists\, \mathcal{T}, \mathcal{A}, \mathcal{O} : |Sol_{\mathbf{cRAP}}| = 2^{|A|}$

Proof. To simplify notation we denote the dominance relation induced by $\mathbf{cRAP}$ by $\prec^c$, and by $\preceq^s$ its counterpart from $\mathbf{sRAP}$.
For the proof of property a), let $(A, O) \in Sol_{\mathbf{cRAP}}$ and assume $(A, O) \notin Sol_{\mathbf{sRAP}}$. Then there must exist $(A', O') \in Sol_{\mathbf{sRAP}}$ such that $(A', O') \prec^s (A, O)$ which is by Definition 3.1 equivalent to $A' \subset A \wedge O' \supseteq O$ or $A' \subseteq A \wedge O' \supset O$. The former case implies $A' \prec^c A$ and $O' \preceq^c O$ due to (anti-)monotonicity for set inclusion, which gives $(A', O') \prec^c (A, O)$ and leads to a contradiction with the assumption. The second case is falsified similarly, concluding the proof.
Property b) follows directly from a) in combination with Proposition 3.4.
The proof of property c) is identical to the one for Proposition 3.4 d). $\qquad\square$

Again, while the worst-case estimate of $Sol_{\mathbf{RAP}}$ is not necessarily encouraging, the potential for domination to occur is larger than in the subset-based instantiation. Dependent on the nature of the axioms in $\mathcal{T}$, $\mathcal{A}$ and $\mathcal{O}$, practical results may be much better. As before, no runtime evaluation of cardinality-based orders has been conducted due to the fact that the properties of the concrete knowledge bases chosen for evaluation would dominate the results.

Example 3.3 (Comparing nclusion- and cardinality-based orders)
In Example 3.1 which employed inclusion-based dominance, we got two valid non-trivial solutions (A_1, O_1) and (A_2, O_2) defined by:

$$A_1 = \{\mathsf{System} \sqsubseteq \exists\,\mathsf{operatesIn}\,.\,\mathsf{PowerSupplyFluctuations}\}$$
$$O_1 = \{\mathsf{MCU} \sqsubseteq \exists\,\mathsf{shows}\,.\,\mathsf{IntermittentOutages},$$
$$\mathsf{Gripper} \sqsubseteq \exists\,\mathsf{shows}\,.\,\mathsf{FullyFunctional}\}$$
$$A_2 = \{\mathsf{System} \sqsubseteq \exists\,\mathsf{operatesIn}\,.\,\mathsf{ControlSWMalfunction}\}$$
$$O_2 = \{\mathsf{MCU} \sqsubseteq \exists\,\mathsf{shows}\,.\,\mathsf{IntermittentOutages}\}$$

Changing to cardinality-based orders $\preceq_\mathcal{A}$ and $\preceq_\mathcal{O}$ changes this picture: Both solutions require one assumption, but the first one explains two observations as compared to one explanation entailed by the second one. This leads to the second solution being dominated by the first one $((A_1, O_1) \prec (A_2, O_2))$, thereby reducing $Sol_{\mathbf{RAP}}$ to $\{(\emptyset, \emptyset), (A_1, O_1)\}$ (effectively shrinking the number of solutions by one third).

Weight-based Orders

Weight-based (pre)orders attribute a (typically real-valued) weight to each element of $\mathcal{A}$ and $\mathcal{O}$ or, alternatively, to every subset of $\mathcal{A}$ and $\mathcal{O}$; the element-based members of the family can furthermore be distinguished by the method of combining the weights of elements to yield the weight of a set. Weights of sets are compared straightforwardly using the standard $\leq$-operator. We denote this family of instantiations by $\mathbf{RAP} = (\mathcal{T}, \mathcal{A}, \mathcal{O}, \leq^{\varsigma_\mathcal{A}}, \geq^{\varsigma_\mathcal{O}})$ defined by instantiation-specific *combination functions* $\varsigma_\mathcal{A}, \varsigma_\mathcal{O}$ that induce preorders $A \leq^{\varsigma_\mathcal{A}} A'$ if and only if $\varsigma_\mathcal{A}(A) \leq \varsigma_\mathcal{A}(A')$ and $O \geq^{\varsigma_\mathcal{A}} O'$ if and only if $\varsigma_\mathcal{A}(O) \geq \varsigma_\mathcal{A}(O')$.

Different choices of weight domains and combination functions result in a variety of semantics for these numbers: For example combining arbitrary values from $\mathbb{R}$ by addition realizes cost-based semantics, while the combination of domain $\mathbb{R} \cap [0; 1]$ with the max operator defines an analogue to Zadeh semantics for fuzzy logic (Zadeh, 1965). The cardinality-based order defined in the previous section can be seen as a special case of weight-based orders that is defined by a weight of 1 for each assumption / observation and addition as combination function. This choice may however not be made arbitrarily as to not violate the monotonicity requirements: As an example, in all the examples introduced before weights must be non-negative to result in valid instantiations of the framework.

Little can be said about this family of instantiations unless further restrictions are imposed on $\preceq_{\mathcal{A}}$ and $\preceq_{\mathcal{O}}$: If Zadeh style fuzzy semantics are employed and all abducibles (observations) are asserted the same weight $w_{\mathcal{A}}$ ($w_{\mathcal{O}}$), there may be up to $O\left(2^{(|\mathcal{A}|+|\mathcal{O}|)}\right)$ solutions, some of them clearly redundant in a logical sense. For additive weights taken from $\mathbb{R} \cap [0;1]$, any non-zero choice of weights guarantees the same upper bound identified in Proposition 3.8 for the cardinality-based case.

To conclude our general presentation of relaxed abduction, we would like to draw the readers' attention to the fact that not only abduction, but also *axiom pinpointing* as introduced in Baader et al. (2007) can be realised using the framework of relaxed abduction. In fact, relaxed abduction can solve multiple pinpointing problems for the same knowledge base in parallel, as the following proposition shows:

Proposition 3.9 (Relaxed abduction can simulate pinpointing)
Let $\mathcal{T}$ be a consistent knowledge base, and $O = \{o_1, \ldots, o_k\}$ a set of axioms such that $\mathcal{T} \models O$. To determine which subset(s) from $\mathcal{T}$ entail o_i (i. e. to solve the axiom pinpointing problem *for o_i), it suffices to solve* $\mathbf{RAP} = (\emptyset, \mathcal{T}, \mathcal{O}, \subseteq_{\mathcal{T}}, \supseteq_{\mathcal{O}})$. *Then, every solution $(T, \{\ldots, o_i, \ldots\})$ encodes a solution to the axiom pinpointing problem for o_i.*

Having introduced the general framework of relaxed abduction, we now turn to the task of solving relaxed abduction problems in the general case, where the preference relations $\preceq_{\mathcal{A}}$ and $\preceq_{\mathcal{O}}$ are regarded as black boxes and no assumptions on their properties other than the ones required by Definition 3.1 are made. We will return to a deeper analysis of the properties of the preorders employed later on, when we investigate early pruning of suboptimal solution candidates.

3.3 Solving Relaxed Abduction

Solving a relaxed abduction problem $\mathbf{RAP} = (\mathcal{T}, \mathcal{A}, \mathcal{O}, \preceq_{\mathcal{A}}, \preceq_{\mathcal{O}})$ consists of determining exactly the elements of the solution set $Sol_{\mathbf{RAP}}$. From this section on, we focus our attention on knowledge bases expressed using description logics due to the fact that decidability is automatically guaranteed in this framework.

In the following we present two alternative approaches to the task at hand: Our first, generic method determines the solution set by a series of exponentially many calls to a regular deductive DL reasoner. Expressiveness of the description logic is only restricted by the underlying reasoning engine; in particular all features of OWL 2 DL (i. e. of the description logic $\mathcal{SROIQ}$) can be supported when a state-of-the-art DL inference engine is used. We additionally present a solution designed specifically for the description logic $\mathcal{EL}^+$ that is used conveniently for the representation of diagnostic ontologies as shown for example in (Hubauer et al., 2012). As the $\mathcal{EL}$ family of description logics was tailored specifically towards tractable (i. e. PTIME) reasoning, its member $\mathcal{EL}^+$ provides a promising basis for an efficient implementation of relaxed abduction as well. Note however that the latter approach can be extended straightforwardly to other non-polynomial description logics that permit completion-based reasoning as in the $\mathcal{EL}$ family.

For simplicity, we assume in both approaches that the observation set $\mathcal{O}$ contains only axioms expressing inclusion between (potentially complex) concepts, but not between roles. While this restriction significantly simplifies presentation, it did not lead to any restrictions in the use cases motivating this research (see Sections 4.3.1 and 4.3.2 for details). However, the algorithms proposed could easily be extended to also handle role inclusions, given the need by a future use case.

3.3.1 A Generic Solution

It is well known (for instance see the paper by Console et al., 1991) that logic-based abduction can be reduced to deduction by iterating over all subsets $A \subseteq \mathcal{A}$, testing whether A is consistent with $\mathcal{T}$ and $A \models_\mathcal{T} \mathcal{O}$ holds. Obviously, this approach requires an exponential number of calls to the deductive reasoner, one for every subset of $\mathcal{A}$. We extend this approach to the relaxed problem variant in this section. The deductive reasoning component is hereby considered as a black-box system whose internal processing is unknown, the solution developed here is therefore a typical representative of the class of *black-box algorithms*.

Algorithm

Our black-box solution to solving relaxed abduction is outlined in Algorithm 3.1. As entailment checking is delegated to a DL reasoner, this algorithm is capable of solving relaxed abduction formulated over arbitrary description logics. After introducing the algorithm, we prove its termination and correctness in Proposition 3.10.

Algorithm 3.1: RAPsolve-Generic

Input: RAP $= (\mathcal{T}, \mathcal{A}, \mathcal{O}, \preceq_{\mathcal{A}}, \preceq_{\mathcal{O}})$
Output: $Sol_{\mathbf{RAP}}$

```
// solution generation
```
1 Sol $\leftarrow \emptyset$;
2 **foreach** A $\subseteq \mathcal{A}$ **do**
3   **if** $\mathcal{T} \cup$ A $\not\models \bot$ **then**
4   O $\leftarrow \emptyset$;
5   **foreach** o $\in \mathcal{O}$ **do**
6   **if** $\mathcal{T} \cup$ A $\models$ o **then** O $\leftarrow$ O $\cup \{$o$\}$;
7   ;
8   Sol $\leftarrow$ Sol $\cup \{($A, O$)\}$;

```
// solution selection
```
9 **foreach** (A, O) $\in$ Sol **do**
10   **foreach** (A$'$, O$'$) $\in$ Sol **do**
11   **if** (A, O) $\prec$ (A$'$, O$'$) **then** Sol $\leftarrow$ Sol $\setminus \{($A$'$, O$')\}$;
12   ;

13 **return** Sol;

Proposition 3.10 (Correctness of RAPsolve-Generic**)**
Let **RAP** $= (\mathcal{T}, \mathcal{A}, \mathcal{O}, \preceq_{\mathcal{A}}, \preceq_{\mathcal{O}})$ *be a relaxed abduction problem where the preorders* $\preceq_{\mathcal{A}}$ *and* $\preceq_{\mathcal{O}}$ *are effectively computable. The execution of Algorithm 3.1 for* **RAP** *terminates in finite time and returns* $Sol_{\mathbf{RAP}}$.

Proof. To show termination note that the generation loop starting with line 2 is executed a finite number of times ($2^{|\mathcal{A}|}$, actually), and the same holds for its inner loop ($|\mathcal{O}|$ iterations). Entailment for DLs can be determined effectively (this is actually a defining property of this family of languages). By Proposition 3.4 the selection loop starting at line 9 is also executed at most $2^{|\mathcal{A}|}$ times. Again, the basic operations are effective (testing $\prec$ comprises testing $\preceq_{\mathcal{A}}$ and $\preceq_{\mathcal{O}}$ which can be done effectively by requirement). For correctness of Algorithm 3.1, note that the generation loop tests all

subsets of $\mathcal{A}$ and is thus complete. The same holds true for the inner test loop, where Proposition 3.4 a) makes it possible to determine the single permitted set O by stepwise addition of single observations. Furthermore, lines 3 and 6 guarantee that the candidates generated meet the purely logical constraints for a solution. Finally, minimality of the solution set is ensured by the selection loop as all pairs of candidates are compared, and dominated elements are removed in line 11. The remaining elements must therefore be dominance-free and thus $\preceq$-minimal. Taken together with lines 3 and 6 this guarantees the soundness of the result set. □

The formulation of Algorithm 3.1 is optimized for comprehensiveness and for the ease of proving Proposition 3.10. With regards to a practical implementation there are several straightforward modifications to this algorithm suited to improve its runtime: Firstly, for typical reasoners the multitude of entailment checks in lines 3 and 6 can be realized more efficiently by classifying $\mathcal{T} \cup A$ once and testing the entailments via lookups in the classified knowledge base. It is clear to see that this affects neither termination nor correctness. Moreover, information reuse can be maximized by iterating over the subsets of $\mathcal{A}$ in order of increasing size:

- If $A \subseteq \mathcal{A}$ is found to be inconsistent with $\mathcal{T}$ in line 3, then all supersets of A may be pruned in the outer loop (line 2).
- Entailment tests in the inner loop (line 6) can be saved by initializing O with $\bigcup_{A' \subseteq A \wedge (A',O') \in Sol} O'$ instead of $\emptyset$ (line 4), and testing entailment only for elements from $\mathcal{O} \setminus O$.
- If the deductive reasoner employed supports incremental reasoning, the total time required for classifying the different $\mathcal{T} \cup A$ can be further reduced by e.g. classifying $\mathcal{T}$ once and adding the axioms of the sets A incrementally, for example using one copy per chain of subsets.

Is it easy to see that termination and correctness are retained if one or all of these modifications are employed. Since neither of the modifications depends on the choice of $\preceq_{\mathcal{A}}$ and $\preceq_{\mathcal{O}}$, they can safely be adapted in all concrete instantiations.

Complexity Analysis

Three non-elementary operations are repeatedly carried out by our black-box algorithm for solving relaxed abduction: the check for $\prec$-minimality in line 11, and the consistency- respectively entailment-tests in the generation loop. As motivated before, the latter two can both be realized by classifying the knowledge base in question. The overall runtime of Algorithm 3.1 therefore crucially depends on the complexity of classifying a L-knowledge base (L being the DL unifying the languages used to express $\mathcal{T}$, $\mathcal{A}$, and $\mathcal{O}$), and on the time required to determine whether one solution dominates another one. In Proposition 3.11 we determine a bound on its runtime that takes these dependencies into account.

Proposition 3.11 (Complexity of RAPsolve-Generic)
*Let L-**RAP** $= (\mathcal{T}, \mathcal{A}, \mathcal{O}, \preceq_{\mathcal{A}}, \preceq_{\mathcal{O}})$ be a relaxed abduction problem, $T_L^{class}(n)$ an upper bound on the complexity of classifying a L-knowledge base of size n, and $T_{\prec_{\mathcal{A}}}(n)$ (or $T_{\prec_{\mathcal{O}}}(n)$, respectively) be upper bounds on the complexity of comparing two sets of maximum size n w. r. t. $\prec_{\mathcal{A}}$ ($\prec_{\mathcal{O}}$). Then $T_{RAPsolve-Generic}(\mathbf{RAP})$, the runtime for solving **RAP** using Algorithm 3.1, is given by*

$$O\left(2^{|\mathcal{A}|} \cdot \left(T_L^{class}(|\mathcal{T} \cup \mathcal{A}|) + |\mathcal{O}|\right) + 4^{|\mathcal{A}|} \cdot \left(T_{\prec_{\mathcal{A}}}(|\mathcal{A}|) + T_{\prec_{\mathcal{O}}}(|\mathcal{O}|)\right)\right).$$

Proof. Follows directly from the pseudocode of Algorithm 3.1 and Proposition 3.4 c). □

Note that a tighter bound could be derived by using the complexity of a single entailment test $T_L^{\models}(n)$ instead of $T_L^{class}(n)$. However, both complexities are typically in the same general complexity class (e. g. for polynomial entailment testing, classification is also polynomial). As complexity results for classification are typically more established, we choose this as a baseline.

Proposition 3.11 shows that the effort for comparing solution candidates in the selection loop can take up a considerable part of the execution time for description logics where classification is tractable, while classification complexity will dominate the overall runtime for more complex DLs. In particular, if classification for L is C-complete, then RAPsolve-Generic is in ExpTime^C (i. e. ExpTime with C-oracle) if C is not easier than ExpTime, and $T_{\prec}$ is small (e. g. linear). Note that the worst-case complexity derived in 3.11 is retained even if the optimizations discussed before are employed.

The generic algorithm investigated in this section serves two purposes: Primarily, it provides a method for solving relaxed abduction problems without any assumptions on the underlying description logic. Moreover, this generic method can serve as a baseline for the evaluation of more sophisticated implementations as presented next. It is worthwhile to point out the close relationship between this black-box realisation of relaxed abduction and the notion of model-based diagnosis (De Kleer & Kurien, 2003), where assumptions on the abnormality (and, potentially, the normality) of system components serve as explanations for the occurrence of certain observations. The iteration over subsets of $\mathcal{A}$ can then be understood as testing different potential solutions. Similarly to logic-based abduction (and different from the approach taken in this thesis), a solution in the context of model-based diagnosis must be consistent with all observations made. The same holds true for the extension of model-based diagnosis to different fault modes that lead to different behaviour of the faulty component. Another closely connected approach is qualitative reasoning and its application to model-based diagnostics (Struss, 1997), where the quantitative behaviour of complex systems is abstracted by qualitative relations. Based on these methods, an approach similar to model-based diagnosis can then be used to explain the observations about such a system.

3.3.2 A Completion-based Algorithm for $\mathcal{EL}^+$

Algorithm RAP-Generic introduced in the preceding section can be understood to perform an uninformed search through the space of pairs (A, O), without making use of the internal structure of the problem as defined by $\mathcal{T}$. We now present an algorithm designed specifically for the description logic $\mathcal{EL}^+$ that uses the special structure of derivations in this description logic to guide the search process. To make this possible, we add as a technical requirement for the $\mathcal{EL}^+$-specific algorithm that the set of observations does not include any role inclusion axioms.[6]

Consequence-driven Classification of $\mathcal{EL}^+$ Knowledge Bases

The description logic $\mathcal{EL}^+$ belongs to a family of DLs for which standard reasoning tasks such as the classification of a knowledge base can be accomplished by means of *consequence-driven reasoning*. In a nutshell, consequence-driven reasoning starts with a set of trivially true axioms (tautologies such as $A \sqsubseteq \top$ and $A \sqsubseteq A$) and uses a set of completion rules specific for the logic at hand to derive new valid consequences from $\mathcal{T}$ in the style of Gentzen's sequent calculus (Gentzen, 1934, 1935). The process terminates when no new consequences can be derived. This approach differs fundamentally from the refutation-based method employed in tableaux-based reasoners,

---

[6]While this restriction is required for technical reasons, it does not negatively affect the applicability of the approach to diagnostics: Here, observations represent relations between (classes of) entities. Role inclusion axioms, in contrast, express that any *potential* pair of entities related by the superproperty must also be related via the subpropery - which cannot even be concluded safely from (typically incomplete) observations.

(NR1-1)	$r_1 \circ \cdots \circ r_k \sqsubseteq s$	$\longrightarrow$	$\{r_1 \circ \cdots \circ r_{k-1} \sqsubseteq u, u \circ r_k \sqsubseteq s\}$
(NR1-2)	$C \sqcap \tilde{D} \sqsubseteq E$	$\dashrightarrow$	$\{\tilde{D} \sqsubseteq A, C \sqcap A \sqsubseteq E\}$
(NR1-3)	$\exists r. \tilde{C} \sqsubseteq D$	$\longrightarrow$	$\{\tilde{C} \sqsubseteq A, \exists r. A \sqsubseteq D\}$
(NR2-1)	$\tilde{C} \sqsubseteq \tilde{D}$	$\longrightarrow$	$\{\tilde{C} \sqsubseteq A, A \sqsubseteq \tilde{D}\}$
(NR2-2)	$B \sqsubseteq \exists r. \tilde{C}$	$\longrightarrow$	$\{B \sqsubseteq \exists r. A, A \sqsubseteq \tilde{C}\}$
(NR2-3)	$B \sqsubseteq C \sqcap D$	$\longrightarrow$	$\{B \sqsubseteq C, B \sqsubseteq D\}$

Figure 3.4: Normalisation rules for $\mathcal{EL}^+$

where a subsumption $A \sqsubseteq B$ is shown to be entailed by a knowledge base $\mathcal{T}$ by proving that adding the negation $A \sqcap \neg B$ leads to a contradiction, i.e. $\mathcal{T} \cup \{(A \sqcap \neg B)(a)\} \models \bot$ for a fresh constant a.

Following the approach of Baader (2003); Baader et al. (2005a), the first step in using consequence-driven reasoning for the classification of a knowledge base $\mathcal{T}$ consists of *normalizing* $\mathcal{T}$ into a so-called conservative extension[7] $\mathcal{T}'$ that only contains axioms of restricted syntactical form. For $\mathcal{EL}^+$, it is known that any knowledge base can be transformed to contain only axioms of the following types (where $C_1, C_2, D \in N_C^\top$ and $r_1, r_2, s \in N_R$): **(NF1)** $C_1 \sqsubseteq D$, **(NF2)** $C_1 \sqcap C_2 \sqsubseteq D$, **(NF3)** $C_1 \sqsubseteq \exists r_1 . C_2$, **(NF4)** $\exists r_1 . C_2 \sqsubseteq D$, **(NF5)** $r_1 \sqsubseteq s$, and **(NF6)** $r_1 \circ r_2 \sqsubseteq s$. This can be accomplished in linear time in the size of $\mathcal{T}$ and with a linear increase in size using the rules depicted in Figure 3.4 (adapted from Baader et al., 2005b), where $\tilde{C}$, $\tilde{D}$ denote complex concepts, A is a new concept name, u is a new role name, and all other concept/role names are unrestricted). Note that due to the introduction of fresh names for auxiliary concepts and roles no unwanted additional conclusions may occur, i.e. soundness is guaranteed.

This normalised ontology then serves as input for the completion algorithm based on the rules shown in Figure 3.5. These rules can be understood as follows: If the premises above the horizontal line are already known to be consequences of the knowledge base $\mathcal{T}$, and the axiom given in square brackets is contained in $\mathcal{T}$, then the conclusion below the horizontal line is a valid consequence of $\mathcal{T}$ as well. Note that there is exactly one completion rule for each type of normal form axiom, plus two initialization rules that provide the tautologies that serve as starting points for the completion process. The applicability of completion rule **(CRi)** is guarded by the presence of an appropriate **(NFi)**-axiom in $\mathcal{T}$, i.e. the information expressed in the knowledge base $\mathcal{T}$ determines the flow of rule applications and thus the

---

[7]$\mathcal{T}'$ is a conservative extension of $\mathcal{T}$ if the signature of $\mathcal{T}'$ is a superset of the signature of $\mathcal{T}$, and the restriction of $\mathcal{T}'$ to the signature of $\mathcal{T}$ is logically equivalent to $\mathcal{T}$.

$$(\text{IR1}) \ \overline{\mathsf{C} \sqsubseteq \mathsf{C}} \qquad (\text{IR2}) \ \overline{\mathsf{C} \sqsubseteq \top}$$

$$(\text{CR1}) \ \frac{\mathsf{C} \sqsubseteq \mathsf{C}_1}{\mathsf{C} \sqsubseteq \mathsf{D}} \ [\mathsf{C}_1 \sqsubseteq \mathsf{D} \in \mathcal{T}]$$

$$(\text{CR2}) \ \frac{\mathsf{C} \sqsubseteq \mathsf{C}_1 \qquad \mathsf{C} \sqsubseteq \mathsf{C}_2}{\mathsf{C} \sqsubseteq \mathsf{D}} \ [\mathsf{C}_1 \sqcap \mathsf{C}_2 \sqsubseteq \mathsf{D} \in \mathcal{T}]$$

$$(\text{CR3}) \ \frac{\mathsf{C} \sqsubseteq \mathsf{C}_1}{\mathsf{C} \sqsubseteq \exists r_1 . \mathsf{C}_2} \ [\mathsf{C}_1 \sqsubseteq \exists r_1 . \mathsf{C}_2 \in \mathcal{T}]$$

$$(\text{CR4}) \ \frac{\mathsf{C} \sqsubseteq \exists r_1 . \mathsf{C}_1 \qquad \mathsf{C}_1 \sqsubseteq \mathsf{C}_2}{\mathsf{C} \sqsubseteq \mathsf{D}} \ [\exists r_1 . \mathsf{C}_2 \sqsubseteq \mathsf{D} \in \mathcal{T}]$$

$$(\text{CR5}) \ \frac{\mathsf{C} \sqsubseteq \exists r_1 . \mathsf{D}}{\mathsf{C} \sqsubseteq \exists s . \mathsf{D}} \ [r_1 \sqsubseteq s \in \mathcal{T}]$$

$$(\text{CR6}) \ \frac{\mathsf{C} \sqsubseteq \exists r_1 . \mathsf{C}_1 \qquad \mathsf{C}_1 \sqsubseteq \exists r_2 . \mathsf{D}}{\mathsf{C} \sqsubseteq \exists s . \mathsf{D}} \ [r_1 \circ r_2 \sqsubseteq s \in \mathcal{T}]$$

Figure 3.5: Completion rules for classification in $\mathcal{EL}^+$

resulting derivations. When the completion process terminates (i.e. no more rules are applicable), all valid subsumptions have been derived, and $\mathcal{T}$ is therefore completely classified. This approach is known to be optimal for classifying $\mathcal{EL}^+$ knowledge bases.

The classification (or, rather, saturation) procedure outlined above is typically realized based on two mappings S and R that make the subsumptions derived so far explicit. To understand these mappings, observe that the premises and conclusions of the completion rules only consist of two types of axioms, namely **(NF1)** $\mathsf{C} \sqsubseteq \mathsf{D}$ and **(NF3)** $\mathsf{C} \sqsubseteq \exists r . \mathsf{D}$. A subsumption $\mathsf{C} \sqsubseteq \mathsf{D}$ is then represented as an S-entry $\mathsf{D} \in S(\mathsf{C})$, whereas a subsumption $\mathsf{C} \sqsubseteq \exists r . \mathsf{D}$ corresponds to an R-entry $(\mathsf{C}, \mathsf{D}) \in R(r)$. In accordance with the initialisation rules **(IR1)** and **(IR2)**, S is initialized by $S(\mathsf{C}) := \{\mathsf{C}, \top\}$ for each concept name in $\mathcal{T}$, and $R(r) := \emptyset$ for all role names in $\mathcal{T}$. Testing premises and adding new conclusions then corresponds to the lookup and addition of entries to these mappings. As Baader et al. have shown, this

completion process terminates after polynomially many rule applications;
the final classification result can then be read off the mappings S and R
(Baader et al., 2005b).

Towards Consequence-driven Relaxed Abduction

We now propose a consequence-driven reasoning scheme for solving relaxed
abduction problems. The basic intuition behind our approach is to extend
the consequence-driven classification algorithm with adequate bookkeeping
that makes explicit the information on assumptions required for deriving
a conclusion, and observations explained so far. In this respect, it can be
seen to extend upon previous work on axiom pinpointing in $\mathcal{EL}^+$ (Baader
et al., 2007) where transitions are labelled to create a pinpointing formula,
but also on earlier results on node labels in ATMS (De Kleer, 1986) and
dependency-directed backtracking in tableaux (Freeman, 1995). We present
our results in two consecutive steps: We motivate the solution from the
perspective of derivation (hyper-)graphs, which facilitates presentation and
helps motivate our approach. In the subsequent line we cast these ideas into
an actual algorithm that extends existing algorithms for completion-based
classification in $\mathcal{EL}^+$. Similarly to consequence-driven classification, we
require the input ontologies to be in normal form.

Definition 3.6 (Normal form for relaxed abduction problems)
A relaxed abduction problem **RAP** $= (\mathcal{T}, \mathcal{A}, \mathcal{O}, \preceq_\mathcal{A}, \preceq_\mathcal{O})$ *is in* normal form
if and only if the knowledge bases $\mathcal{T}$, $\mathcal{A}$ *and* $\mathcal{O}$ *are in normal form.*

A straightforward approach to normalising an $\mathcal{EL}^+$-**RAP** consists of independently transforming its three knowledge bases into normal form. This
solution is however problematic due to the fact that one axiom in the original
knowledge base corresponds to several axioms after normalisation. Therefore,
for one observation to be entailed, we would possibly have to track the entailment status of a set of normal form axioms. A similar problem occurs with
assumptions, where assuming one axiom from $\mathcal{A}$ then corresponds to adding
several normal form axioms. To avoid these complications, we have chosen
a slightly different approach to normalising a relaxed abduction problem
shown in Algorithm 3.2, where `normalise` is a function that normalises an
axiom set using the rules given in Figure 3.4.

Basically, our algorithm for normalising a relaxed abduction problem
treats $\mathcal{T}$ as usual, whereas non-normal axioms from $\mathcal{A}$ (or $\mathcal{O}$) are replaced
by a normal-form proxy axiom in $\mathcal{A}$' ($\mathcal{O}$'), and additional normalised axioms
ensuring the intended semantics are added to $\mathcal{T}'$. During this process, the

---

Algorithm 3.2: RAPnorm

Input: RAP $= (\mathcal{T}, \mathcal{A}, \mathcal{O}, \preceq_{\mathcal{A}}, \preceq_{\mathcal{O}})$ *– an* $\mathcal{EL}^+$*-***RAP** *over* $N_C^\top$ *and* N_R
Output: RAP$' = (\mathcal{T}', \mathcal{A}', \mathcal{O}', \preceq'_{\mathcal{A}}, \preceq'_{\mathcal{O}})$ *– the normalised input* **RAP**
$\quad$ M *– the (de)normalisation mapping*

$\quad$ // Normalise $\mathcal{T}$ as usual
1 $\mathcal{T}' \leftarrow$ normalise($\mathcal{T}$);
$\quad$ // Special treatment for $\mathcal{A}$ and $\mathcal{O}$ (proxy axioms)
2 $\mathcal{A}' \leftarrow \emptyset$;
3 $\mathcal{O}' \leftarrow \emptyset$;
4 M $\leftarrow \emptyset$;
5 **foreach** ax $\in \mathcal{A}$ **do**
6 $\quad\quad$ **if** ax *is a normal form axiom* **then** $\mathcal{A}' \leftarrow \mathcal{A}' \cup \{\text{ax}\}$;
7 $\quad\quad$ **else**
8 $\quad\quad\quad$ **if** ax *is a concept inclusion axiom* $\tilde{C} \sqsubseteq \tilde{D}$ **then**
9 $\quad\quad\quad\quad$ $\mathcal{A}' \leftarrow \mathcal{A}' \cup \{C^* \sqsubseteq D^*\};$ $\quad$ // C^*, D^* fresh concept names
10 $\quad\quad\quad\quad$ $\mathcal{T}' \leftarrow \mathcal{T}' \cup$ normalise($\{\tilde{C} \sqsubseteq C^*, D^* \sqsubseteq \tilde{D}\}$);
11 $\quad\quad\quad\quad$ M $\leftarrow$ M $\cup \{(C^* \sqsubseteq D^*, \tilde{C} \sqsubseteq \tilde{D})\};$
12 $\quad\quad\quad$ **else if** ax *is a role inclusion axiom* $r_1 \circ \cdots \circ r_m \sqsubseteq s_1 \circ \cdots \circ s_n$ **then**
13 $\quad\quad\quad\quad$ $\mathcal{A}' \leftarrow \mathcal{A}' \cup \{r^* \sqsubseteq s^*\};$ $\quad$ // r^*, s^* fresh role names
14 $\quad\quad\quad\quad$ $\mathcal{T}' \leftarrow \mathcal{T}' \cup$ normalise($\{r_1 \circ \cdots \circ r_m \sqsubseteq r^*, s^* \sqsubseteq s_1 \circ \cdots \circ s_n\}$);
15 $\quad\quad\quad\quad$ M $\leftarrow$ M $\cup \{(r^* \sqsubseteq s^*, r_1 \circ \cdots \circ r_m \sqsubseteq s_1 \circ \cdots \circ s_n)\};$
16 **foreach** ax $\in \mathcal{O}$ **do**
17 $\quad\quad$ **if** typeof(ax) $=$ *(NF1)* *or* typeof(ax) $=$ *(NF3)* **then**
$\quad\quad\quad$ $\mathcal{O}' \leftarrow \mathcal{O}' \cup \{\text{ax}\}$;
18 $\quad\quad$ **else if** ax *is a concept inclusion axiom* $\tilde{C} \sqsubseteq \tilde{D}$ **then**
19 $\quad\quad\quad$ $\mathcal{O}' \leftarrow \mathcal{O}' \cup \{C^* \sqsubseteq D^*\};$ $\quad$ // C^*, D^* fresh concept names
20 $\quad\quad\quad$ $\mathcal{T}' \leftarrow \mathcal{T}' \cup$ normalise($\{C^* \sqsubseteq \tilde{C}, \tilde{D} \sqsubseteq D^*\}$);
21 $\quad\quad\quad$ M $\leftarrow$ M $\cup \{(C^* \sqsubseteq D^*, \tilde{C} \sqsubseteq \tilde{D})\};$

$\quad$ // Extend preference relations to proxy axioms
22 $\preceq'_{\mathcal{A}} \leftarrow \preceq_{\mathcal{A}} \cup \{(A'_1, A'_2) \mid \exists A_1, A_2 \subseteq \mathcal{A} . (A_1, A'_1) \in$ M
$\quad \wedge (A_2, A'_2) \in$ M $\wedge A_1 \preceq_{\mathcal{A}} A_2\};$
23 $\preceq'_{\mathcal{O}} \leftarrow \preceq_{\mathcal{O}} \cup \{(O'_1, O'_2) \mid \exists O_1, O_2 \subseteq \mathcal{O}. (O_1, O'_1) \in$ M
$\quad \wedge (O_2, O'_2) \in$ M $\wedge O_1 \preceq_{\mathcal{O}} O_2\};$
$\quad$ // Return normalisation result
24 **return** (**RAP**$'$, M);

---

algorithm constructs a mapping $\mathbf{M}$ that stores which proxy axiom in $\mathcal{A}'$ (or $\mathcal{O}'$) represents which original assumption (or observation). To keep the mapping as small as possible, axioms that already are in normal form are considered their own proxy and not added to $\mathbf{M}$. Since this mapping is bijective[8], if can be used to de-normalized solutions easily when the algorithm has completed. In addition, $\mathbf{M}$ induces two projection functions $\mathbf{M}_1^\pi$ and $\mathbf{M}_2^\pi$ that can be seen to project a $\mathbf{M}$-entry on its first (second) component, based on a lookup on the respective other dimension. That is, $\mathbf{M}_2^\pi(A)$ maps A to its normalization A' while $\mathbf{M}_1^\pi(A')$ maps the normalized axiom back to the original one.

Note that this procedure leaves the cardinalities of $\mathcal{A}$ and $\mathcal{O}$ unchanged, thereby avoiding the problem outlined before as well as any blow-up in the labels. As shown in the following proposition, this modified normalisation procedure retains correctness. From here on, we assume that abduction problems are normalised using Algorithm 3.2.

Proposition 3.12 (Correctness of RAPnorm)
Let $\mathbf{RAP} = (\mathcal{T}, \mathcal{A}, \mathcal{O}, \preceq_\mathcal{A}, \preceq_\mathcal{O})$ *be an* $\mathcal{EL}^+$ *relaxed abduction problem, and* $\mathbf{RAP}'$ *the corresponding normalised problem, with mapping* $\mathbf{M}$ *created during normalisation. Slightly abusing notation, we extend* $\mathbf{M}$ *to the structure of a solution set (i. e. a set of pairs of sets of axioms).[9] Then,* $Sol_{\mathbf{RAP}'} = \mathbf{M}_2^\pi(Sol_{\mathbf{RAP}})$.

Proof. First, note that the normalisation step encapsulated in the `normalise` function is known to be entailment-preserving (see Baader et al., 2005b), and $Sol_{\mathbf{RAP}}$ therefore remains unaffected by the normalisation of $\mathcal{T}$.
Regarding the normalisation of $\mathcal{A}$ (loop from line 5 onwards), axioms already in normal form obviously do not affect correctness since they are just copied to $\mathcal{A}'$. In case of an unnormalised concept inclusion axiom $\tilde{\mathsf{C}} \sqsubseteq \tilde{\mathsf{D}}$, it obviously holds that $\{\mathsf{C}^* \sqsubseteq \mathsf{D}^*, \tilde{\mathsf{C}} \sqsubseteq \mathsf{C}^*, \mathsf{D}^* \sqsubseteq \tilde{\mathsf{D}}\} \models \tilde{\mathsf{C}} \sqsubseteq \tilde{\mathsf{D}}$, and thus also $\mathcal{T}' \cup \{\mathsf{C}^* \sqsubseteq \mathsf{D}^*\} \models \tilde{\mathsf{C}} \sqsubseteq \tilde{\mathsf{D}}$. Any conclusion of $\mathcal{T} \cup \{\tilde{\mathsf{C}} \sqsubseteq \tilde{\mathsf{D}}\}$ is therefore also a conclusion of $\mathcal{T}' \cup \{\mathsf{C}^* \sqsubseteq \mathsf{D}^*\}$. Conversely, since the concept names introduced in this step are fresh, no unwanted entailments from $\mathsf{C}^* \sqsubseteq \mathsf{D}^*$ can occur, i. e. the two sets are identical. A similar argument holds for the case of an unnormalised role inclusion axiom. To sum up, the normalisation of $\mathcal{A}$ leaves entailments w. r. t. to the original signature unchanged.

---

[8]The reason for the bijectiveness of $\mathbf{M}$ is that newly created proxy axioms contain unique fresh names, and that each original axiom is translated exactly once.

[9]$\mathbf{M}$ applied to a set of axioms returns the set of mapped axioms. A pair of sets is mapped component-wise. Finally, a set of fairs is mapped on a per element basis, giving a set of mapped pairs.

For the normalisation of $\mathcal{O}$ (loop from line 16 onwards), recall that axioms in $\mathcal{O}$ may not be role inclusion axioms by requirement. It therefore remains to show that $S \models \tilde{C} \sqsubseteq \tilde{D}$ if and only if $S \cup \{C^* \sqsubseteq \tilde{C}, \tilde{D} \sqsubseteq D^*\} \models C^* \sqsubseteq D^*$ (like before, we are trivially done if the axiom is in normal form already). Note that $\{\tilde{C} \sqsubseteq \tilde{D}, C^* \sqsubseteq \tilde{C}, \tilde{D} \sqsubseteq D^*\} \models C^* \sqsubseteq D^*$. Therefore, for any observation entailed by S, its proxy is entailed by $S \cup \{C^* \sqsubseteq \tilde{C}, \tilde{D} \sqsubseteq D^*\}$.

Additionally, due to the introduction of fresh concept names no other, unwanted entailments can occur, justifying the equality of the two sets and concluding the proof. □

Example 3.4 (Normalization)
To illustrate the algorithm, let us have a look at how the abducible axiom

$$\text{System} \sqsubseteq \exists \text{operatesIn} \,.\, \text{PowerSupplyFluctuations}$$

is normalized. Firstly, note that the axiom is not in normal form. Therefore, it is replaced by a new $\mathcal{A}$-axiom

$$\text{System}' \sqsubseteq \text{ExOpInPowerSupplyFluct.}$$

*In this process, two fresh concept names are introduced. $\mathcal{T}$ is extended with the following additional axioms without further normalization, as the axioms are already (**NF1**) and (**NF3**):*

$$\text{System} \sqsubseteq \text{System}'$$
$$\text{ExOpInPowerSupplyFluct} \sqsubseteq \exists \text{operatesIn} \,.\, \text{PowerSupplyFluctuations}$$

Similarly, the mapping set M is extended with the pair having forst compon-ent System $\sqsubseteq \exists$ operatesIn . PowerSupplyFluctuations *and second component* System$'$ $\sqsubseteq$ ExOpInPowerSupplyFluct. *The mapping allows for a direct lookup between original and translated axioms.*

The completion rules depicted in Figure 3.5 can be understood as "con-struction patterns" for a hypergraph whose edges represent the dependencies between subsumptions entailed by the normal form knowledge base $\mathcal{T}$: if a subsumption s can be derived from s_1 and s_2 using one of the completion rules **(CRi)** and an axiom $ax \in \mathcal{T}$, then there exists an edge from s_1 and s_2 to s. We denote this derivation step by the shorthand $(\mathbf{CR}i), ax \vdash \{s_1, s_2\} \mapsto s$. Since the rule set is known to be sound and complete for deduction, a know-ledge base $\mathcal{T}$ can equivalently be represented as a hypergraph constructed this way. This correspondence carries over from classification to relaxed abduction as follows: The $\mathcal{EL}^+$-normal form axioms in $\mathcal{T}$ and $\mathcal{A}$ can justify single derivation steps according to the extended set of completion rules depicted in Figure 3.6 that now contains rules for $\mathcal{T}$- and $\mathcal{A}$-justified edges. The edges of the hypergraph are attributed with weights expressing the relevant bookkeeping information. Finally, elements from $\mathcal{O}$ represent information to be justified (derived), they therefore correspond to vertices of the hypergraph. As mentioned previously, this implies that observation axioms may be of type **(NF1)** and **(NF3)** only, which is ensured in the normalisation process. The intuition of a hypergraph built from **RAP** motivated here is formalized in Definition 3.7 (for a general definition of hypergraphs see Definition 2.6).

Definition 3.7 (Induced hypergraph $\mathcal{H}_{\mathbf{RAP}}$)
Let **RAP** $= (\mathcal{T}, \mathcal{A}, \mathcal{O}, \preceq_{\mathcal{A}}, \preceq_{\mathcal{O}})$ *be a relaxed abduction problem whose $\mathcal{EL}^+$ knowledge bases use only concept names from $N_C^\top$ and role names from N_R. Let $\mathcal{CR}$ denote the set of completion rules depicted in Figure 3.6. The induced hypergraph $\mathcal{H}_{\mathbf{RAP}}$ is a weighted hypergraph $\mathcal{H}_{\mathbf{RAP}} = (V, E)$ defined by*

- $V = \{(\mathsf{C} \sqsubseteq \mathsf{D}), (\mathsf{C} \sqsubseteq \exists \mathsf{r}.\mathsf{D}) \mid \mathsf{C}, \mathsf{D} \in N_C^\top, \mathsf{r} \in N_R\}$
- $E = E^\mathcal{T} \cup E^\mathcal{A}$, *where:*
    - $E^\mathcal{T} = \{(T, h, \{w\}) \in \mathcal{P}(V) \times V \times \mathcal{P}(\mathcal{P}(\mathcal{A}) \times \mathcal{P}(\mathcal{O})) \mid \exists (\boldsymbol{CRi^\mathcal{T}}) \in \mathcal{CR}, \exists ax \in \mathcal{T} : (\boldsymbol{CRi^\mathcal{T}}), ax \vdash T \mapsto h \wedge w = (\emptyset, \{h\} \cap \mathcal{O})\}$
    - $E^\mathcal{A} = \{(T, h, \{w\}) \in \mathcal{P}(V) \times V \times (\mathcal{P}(\mathcal{A}) \times \mathcal{P}(\mathcal{O})) \mid \exists (\boldsymbol{CRi^\mathcal{A}}) \in \mathcal{CR}, \exists ax \in \mathcal{A} : (\boldsymbol{CRi^\mathcal{A}}), ax \vdash T \mapsto h \wedge w = (\{ax\}, \{h\} \cap \mathcal{O})\}$

- $(W, \otimes)$, *the weight system of $\mathcal{H}_{\mathbf{RAP}}$, is defined by $W = \mathcal{P}(\mathcal{P}(\mathcal{A}) \times \mathcal{P}(\mathcal{O}))$ and $L_1 \otimes L_2 = \{(A_1 \cup A_2, O_1 \cup O_2) \mid (A_1, O_1) \in L_1 \wedge (A_2, O_2) \in L_2\}$*

The set of leaves *of $\mathcal{H}_{\mathbf{RAP}}$ is defined by $V \supseteq V_\top = \{(\mathsf{C} \sqsubseteq \mathsf{C}), (\mathsf{C} \sqsubseteq \top) \mid \mathsf{C} \in N_C^\top\}$. Moreover, we denote by $E^\mathcal{A} \supseteq E^{\mathcal{A}|A} = \{(T, h, \{(\{ax\}, O)\}) \in E^\mathcal{A} \mid ax \in A\}$ the* restriction *of $E^\mathcal{A}$ to justifications from $A \subseteq \mathcal{A}$.*

The sets $E^\mathcal{T}$ and $E^\mathcal{A}$ represent derivations justified by a (regular) axiom from $\mathcal{T}$ and derivations requiring an assumption, respectively. This is reflected in the first component of the weight $w = (A, O)$, whereas the second component uniformly stores whether this derivation explains an observation (encoded by $O \neq \emptyset$). The weight space W comprises all sets of tuples (A, O), i.e. all possible labels, the operator $\otimes$ implements conjunction for labels: following the intuition that each entry of L_1 and L_2 represents one possible path, the cojoined path represented by L may use any combination composed of one path from each L_1 and L_2. As can easily be seen, $(W, \otimes)$ is

$$\textbf{(IR1)} \ \frac{}{C \sqsubseteq C} \qquad \textbf{(IR2)} \ \frac{}{C \sqsubseteq \top}$$

$$\textbf{(CR1}^{\mathcal{T}}\textbf{)} \ \frac{C \sqsubseteq C_1}{C \sqsubseteq D} \ [C_1 \sqsubseteq D \in \mathcal{T}] \qquad \textbf{(CR1}^{\mathcal{A}}\textbf{)} \ \frac{C \sqsubseteq C_1}{C \sqsubseteq D} \ [C_1 \sqsubseteq D \in \mathcal{A}]$$

$$\textbf{(CR2}^{\mathcal{T}}\textbf{)} \ \frac{C \sqsubseteq C_1 \qquad C \sqsubseteq C_2}{C \sqsubseteq D} \ [C_1 \sqcap C_2 \sqsubseteq D \in \mathcal{T}]$$

$$\textbf{(CR2}^{\mathcal{A}}\textbf{)} \ \frac{C \sqsubseteq C_1 \qquad C \sqsubseteq C_2}{C \sqsubseteq D} \ [C_1 \sqcap C_2 \sqsubseteq D \in \mathcal{A}]$$

$$\textbf{(CR3}^{\mathcal{T}}\textbf{)} \ \frac{C \sqsubseteq C_1}{C \sqsubseteq \exists r_1 . C_2} \ [C_1 \sqsubseteq \exists r_1 . C_2 \in \mathcal{T}]$$

$$\textbf{(CR3}^{\mathcal{A}}\textbf{)} \ \frac{C \sqsubseteq C_1}{C \sqsubseteq \exists r_1 . C_2} \ [C_1 \sqsubseteq \exists r_1 . C_2 \in \mathcal{A}]$$

$$\textbf{(CR4}^{\mathcal{T}}\textbf{)} \ \frac{C \sqsubseteq \exists r_1 . C_1 \qquad C_1 \sqsubseteq C_2}{C \sqsubseteq D} \ [\exists r_1 . C_2 \sqsubseteq D \in \mathcal{T}]$$

$$\textbf{(CR4}^{\mathcal{A}}\textbf{)} \ \frac{C \sqsubseteq \exists r_1 . C_1 \qquad C_1 \sqsubseteq C_2}{C \sqsubseteq D} \ [\exists r_1 . C_2 \sqsubseteq D \in \mathcal{A}]$$

$$\textbf{(CR5}^{\mathcal{T}}\textbf{)} \ \frac{C \sqsubseteq \exists r_1 . D}{C \sqsubseteq \exists s . D} \ [r_1 \sqsubseteq s \in \mathcal{T}] \qquad \textbf{(CR5}^{\mathcal{A}}\textbf{)} \ \frac{C \sqsubseteq \exists r_1 . D}{C \sqsubseteq \exists s . D} \ [r_1 \sqsubseteq s \in \mathcal{A}]$$

$$\textbf{(CR6}^{\mathcal{T}}\textbf{)} \ \frac{C \sqsubseteq \exists r_1 . C_1 \qquad C_1 \sqsubseteq \exists r_2 . D}{C \sqsubseteq \exists s . D} \ [r_1 \circ r_2 \sqsubseteq s \in \mathcal{T}]$$

$$\textbf{(CR6}^{\mathcal{A}}\textbf{)} \ \frac{C \sqsubseteq \exists r_1 . C_1 \qquad C_1 \sqsubseteq \exists r_2 . D}{C \sqsubseteq \exists s . D} \ [r_1 \circ r_2 \sqsubseteq s \in \mathcal{A}]$$

Figure 3.6: Completion rules for relaxed abduction in $\mathcal{EL}^+$

indeed a commutative monoid. Obviously, it is moreover sufficient to consider simple (i. e. loop-free) hyperpaths in $\mathcal{H}_{\mathbf{RAP}}$ by construction: Since the edges in the graph represent instantiations of the completion rules introduced before, and any such rule is only applied once on the same set of premises. The following proposition reveals a close connection between derivations and hyperpaths in $\mathcal{H}_{\mathbf{RAP}}$.

Proposition 3.13 (Hyperpaths correspond to derivations)
Let $\mathbf{RAP} = (\mathcal{T}, \mathcal{A}, \mathcal{O}, \preceq_A, \preceq_{\mathcal{O}})$ be a relaxed abduction problem, and $\mathcal{H}_{\mathbf{RAP}}$ its induced hypergraph. Moreover, let $A \subseteq \mathcal{A}$, $o_i \in \mathcal{O}$. Then:

a) $(A, \{o_i\})$ is a pre-solution (i. e. $\mathcal{T} \cup A \models o_i$) if and only if there exists an hyperpath $p_{V_\top, o_i}$ in $\mathcal{H}_{\mathbf{RAP}}$ such that $E_{V_\top, o_i} \subseteq E^{\mathcal{T}} \cup E^{\mathcal{A}|A}$.

b) (A, O) is a pre-solution (i. e. $\mathcal{T} \cup A \models O$) if and only if there exists an hyperpath $p_{V_\top, O}$ in $\mathcal{H}_{\mathbf{RAP}}$ such that $E_{V_\top, O} \subseteq E^{\mathcal{T}} \cup E^{\mathcal{A}|A}$.

Proof. We show a) by induction over the number n of derivation steps, respectively the number of edges of $p_{V_\top, o_i}$ (also called the length of the path). If o_i can be shown in $n = 0$ steps, it must be a tautology, i. e. $\emptyset \models o_i$. Then also $o_i \in V_\top$, therefore $p_{V_\top, o_i} = (V_\top, \emptyset)$ by Definition 2.8 and $\emptyset \subseteq E^{\mathcal{T}}$. Conversely if $p_{V_\top, o_i} = (V_\top, \emptyset)$ is a path in $\mathcal{H}_{\mathbf{RAP}}$ then $o_i \in V_\top$, which yields $\emptyset \models o_i$ and $\mathcal{T} \cup A \models o_i$ by left-weakening (c.f. Buss, 1998). Now assume (IH) that a) holds for derivations of length n, and let the derivation require $n + 1$ steps. For the "if" part let $p_{V_\top, o_i}$ be a hyperpath in $\mathcal{H}_{\mathbf{RAP}}$ with $E_{V_\top, o_i} \subseteq E^{\mathcal{T}} \cup E^{\mathcal{A}|A}$. By Definition 2.8, $E_{V_\top, o_i}$ must contain an edge $e^* = (T^*, o_i, w^*)$, and each $t \in T^*$ is reachable by a path using only edges from $E_{V_\top, o_i} \setminus \{e^*\}$. By (IH) $\forall t \in T^* : \mathcal{T} \cup A \models t$, and e^* represents a derivation step $T^* \vdash_{\mathcal{T} \cup A} o_i$ by Definition 3.7. Therefore $\mathcal{T} \cup A \models o_i$. For the "only if" part, let $\mathcal{T} \cup A \models o_i$, and let $T \vdash_{\mathcal{T} \cup A} o_i$ be the final derivation step. By (IH) there exist hyperpaths $p_{V_\top, t}$ for all $t \in T$ with $E_{V_\top, t} \subseteq E^{\mathcal{T}} \cup E^{\mathcal{A}|A}$. The hyperpath $p_{V_\top, o_i}$ can then be constructed by $V_{V_\top, o_i} = \bigcup_{t \in T} V_{V_\top, t}$ and $E_{V_\top, o_i} = \bigcup_{t \in T} E_{V_\top, t}$.
Claim b) follows straightforwardly from a) by taking $V_{V_\top, O} = \bigcup_{o_i \in O} V_{V_\top, o_i}$, $E_{V_\top, O} = \bigcup_{o_i \in O} E_{V_\top, o_i}$, and $p_{V_\top, O} = (V_{V_\top, O}, E_{V_\top, O})$. $\qquad\square$

Solutions (A, O) to $\mathbf{RAP}$ are not only required to represent correct derivations $\mathcal{T} \cup A \models O$, but also be minimal w. r. t. $\preceq$. Intuitively, we are looking for minimum-weight paths in $\mathcal{H}_{\mathbf{RAP}}$. However, the general notion of a path weight given in Definition 2.8 does not correctly compute the weight if tautological observations exist. The reason for this is that tautological observations do not have any incoming edges, the observation will therefore never be added to the path weight. We therefore slightly extend the notion of path weights, and relate solutions to minimal paths w. r. t. the extended notion.

Definition 3.8 (Rooted hyperpath weight)
Let $p_{X,T} = (V_{X,T}, E_{X,T})$ be a hyperpath with $w(p_{X,T}) = \bigotimes_{e \in E_{X,T}} w(e)$. The rooted weight *of $p_{X,T}$ with root R is defined by*

$$w_R(p_{X,T}) = w(p_{X,T}) \otimes \bigotimes_{r \in X \cap R} \{(\emptyset, \{r\})\}$$

$$= w(p_{X,T}) \otimes \{(\emptyset, \{X \cap R\})\}.$$

Proposition 3.14 (Hyperpaths with minimum rooted weight are solutions)
Let $\mathbf{RAP} = (\mathcal{T}, \mathcal{A}, \mathcal{O}, \preceq_\mathcal{A}, \preceq_\mathcal{O})$ be a relaxed abduction problem, and $\mathcal{H}_\mathbf{RAP}$ the corresponding induced hypergraph. Then $(A, O) \models \mathbf{RAP}$ if and only if there exists a hyperpath $p_{V_\top, O}$ in $\mathcal{H}_\mathbf{RAP}$ such that $E_{V_\top} \subseteq E^\mathcal{T} \cup E^{\mathcal{A}|A}$ and the $\mathcal{O}$-rooted weight $w_\mathcal{O}(p_{V_\top, O})$ is $\preceq$-minimal for all paths $p_{V_\top, X}$ in $\mathcal{H}_\mathbf{RAP}$.

Proof. Using Proposition 3.13 b), it remains to show that (A, O) is $\preceq$-minimal if and only if $w_\mathcal{O}(p_{V_\top, O})$ is $\preceq$-minimal. For the "only if" direction, assume $(A', O') \prec (A, O)$, which implies $A' \subseteq A \wedge O' \supseteq O$ by definition of $\prec$ and the monotonicity-requirement, where at least one inclusion is strict. If $A' \subset A$ then there must exist a path $p_{V_\top, O'}$ that uses less assumptions, and therefore has a smaller rooted weight than $p_{V_\top, O}$, contradicting its minimality. The case $O' \supset O$ is similar, as all tautological observations are contained in both O and O' anyhow. For the "if" direction, if there is a path $p_{V_\top, O'}$ with $\prec$-smaller weight than $p_{V_\top, O}$, then by the definition of w this path must either use a subset of the assumptions used by the former, or explain a superset of its observations. This contradicts the $\preceq$-minimality of (A, O). $\square$

This shows that a relaxed abduction problem $\mathbf{RAP}$ can be solved by determining weight-minimal hyperpaths in its induced hypergraph $\mathcal{H}_\mathbf{RAP}$. Next, we propose an algorithm that employs this intuition for solving relaxed abduction problems over $\mathcal{EL}^+$ knowledge bases.

Algorithm

This correspondence is utilised in Algorithm `RAPsolve-EL+` to solve a relaxed abduction problem $\mathbf{RAP}$ by constructing bi-criterion shortest hyperpaths in $\mathcal{H}_{\mathbf{RAP}}$ following a so-called label-correcting approach. In a nutshell, label-correcting approaches for determining shortest paths maintain a list of vertices to be processed, initially set to contain the start vertex of the path which is labelled with the identity element (or "zero") for $\otimes$. Iteratively, one element at a time is removed from the list, all its outgoing edges are relaxed, and the labels of the successor vertices are updated (corrected) accordingly. Vertices whose label has changed are re-inserted into the list. When this list becomes empty, the shortest path can be read off the graph. This method has been realized, for instance, in the well-known Dijkstra's Algorithm (Cherkassky et al., 1996). As Guerriero & Musmanno (2001); Skriver (2000) show, the label-correcting method generalizes well to the problem of finding bi-criterion optimal paths and provides the basis for several of the most efficient algorithms known for this problem. Moreover, the approach extends straightforwardly to hypergraphs with the single modification of re-inserting a vertex into the list whenever *any* of its predecessor nodes has been modified.

For comprehending the pseudocode in Algorithms 3.3 to 3.13, it is important to understand that the algorithm does not represent $\mathcal{H}_{\mathbf{RAP}}$ explicitly. Instead, the (essential part of the) hypergraph is represented compactly by means of two labelled mappings

- $S : N_C^\top \mapsto \mathcal{P}(N_C^\top \times \mathcal{P}(\mathcal{P}(\mathcal{A}) \times \mathcal{P}(\mathcal{O})))$, and
- $R : N_R \mapsto \mathcal{P}(N_C^\top \times N_C^\top \times \mathcal{P}(\mathcal{P}(\mathcal{A}) \times \mathcal{P}(\mathcal{O})))$.

This approach straightforwardly extends the method used for representing the unlabelled deviation graph presented in Baader et al. (2005a) to our setting by enhancing the map entries with labels that record all distinct ways of deriving s (i.e. paths $p_{V_\top,s}$ in $\mathcal{H}_{\mathbf{RAP}}$ with $s \in S$). From this information, the solution set $Sol_{\mathbf{RAP}}$ can ultimately be constructed. Label updates are propagated along the hyperedges by means of two operators called meet and join. The meet operator $\otimes$ conjunctively combines the labels of one or two premise vertices, the justification axiom ax, and the conclusion vertex into a label representing a hyperpath to the conclusion with the final hyperedge being justified by ax. It is realised by Algorithm 3.11, constructing the crossproduct of justifications for the premises in line 1, and integrating information on required assumptions and explained observations in the next two lines. The join operator $\oplus$ simply integrates alternative derivations of the same statement by forming the union of the labels. As it is the case for $\otimes$ we require $(W, \oplus)$ to be a commutative monoid, which is met by its implementation in Algorithm 3.10. The propagation terminates when no more effective label updates occur. Next, the meet-closure call in line 24 of Algorithm 3.3 first generates all possible combinations of solutions, potentially introducing new solutions in this process: consider $\mathcal{T} \cup A_1 \models O_1$ and $\mathcal{T} \cup A_2 \models O_2$, then also $\mathcal{T} \cup (A_1 \cup A_2) \models (O_1 \cup O_2)$. Dominated tuples are removed subsequently by `remove-dominated` (Algorithm 3.13). Proposition 3.15 below proves termination and correctness of this algorithm. Note that, for simplicity of notation, we assume that all elements of **RAP** as well as the mappings S and R are global variables that can be read and modified by the main algorithm as well as by any of its subroutines, while all other variables are local. Moreover, all parameters are passed by reference.

Proposition 3.15 (Termination and correctness of `RAPsolve-EL+`)
Let **RAP** $= (\mathcal{T}, \mathcal{A}, \mathcal{O}, \preceq_{\mathcal{A}}, \preceq_{\mathcal{O}})$ *be an* $\mathcal{EL}^+$*-**RAP** where* $\preceq_{\mathcal{A}}$ *and* $\preceq_{\mathcal{O}}$ *are effectively computable. The execution of Algorithm 3.3 for* **RAP** *eventually terminates with result* $Sol_{\mathbf{RAP}}$.

Proof. For the proof of termination, observe that the basic operations `join` (Algorithm 3.10), `meet` (Algorithm 3.11) and `remove-dominated` (Algorithm 3.13) trivially terminate, as label sets are finite. Since N_{C} and N_{R} are finite, the procedures `applyCRx` (Algorithms 3.4 to 3.9) must then terminate as well. The same is true for `meet-closure` (Algorithm 3.12) since the size of L is bounded and both `meet` and `remove-dominated` terminate. Moreover, `applyCRx` returns true if and only if the new label is a superset of the old one (as `join` and `meet` never delete label entries); since the size of labels is bounded as well as the number of axioms, concept names, and role names, they must therefore finally return `false`. Conclusively, the `repeat` loop in `RAPsolve-EL+` must ultimately terminate; Algorithm 3.3 thus terminates.

Correctness: It is known that label correcting methods provide a sound and complete (and, moreover, near-optimal) method for finding shortest bi-criterion paths in graphs (Skriver, 2000). For the transition from graphs to hypergraphs, the main difference is that edges may have multiple (here: two) source nodes. The updating procedure has been adapted accordingly (method `meet`) to take the semantics of label entries into account (c.f. the previous paragraph). Thus, Algorithm 3.3 constructs hyperpaths[10] in $\mathcal{H}_{\mathbf{RAP}}$ that start in V_{T} and collects them in *Sol*. All dominated hyperpaths are removed in the last line of `meet-closure`. Taken together, all weight-minimal hyperpaths in $\mathcal{H}_{\mathbf{RAP}}$ are determined, these correspond to the solutions to **RAP** by Proposition 3.14. □

---

[10] More concretely, it constructs equivalence classes of hyperpaths, where two paths are considered equal if and only if they solely differ w. r. t. the $E^{\mathcal{T}}$-edges used, but coincide w. r. t. A and O.

Algorithm 3.3: RAPsolve-EL+

Input: RAP $= (\mathcal{T}, \mathcal{A}, \mathcal{O}, \preceq_{\mathcal{A}}, \preceq_{\mathcal{O}})$ – an $\mathcal{EL}^+$-**RAP** over $N_C^\top$ and N_R

Output: $Sol_{\mathbf{RAP}}$

    // initialization

1 **foreach** $r \in N_R$ **do**

2     $\lfloor$ $R(r) \leftarrow \emptyset$;

3 **foreach** $C \in N_C^\top$ **do**

4   **if** $C \sqsubseteq C \in \mathcal{O}$ **then** $S(C) \leftarrow \{C : \{(\emptyset, \{C \sqsubseteq C\})\}\}$;

5   **else** $S(C) \leftarrow \{C : \{(\emptyset, \emptyset)\}\}$;

6   **if** $C \sqsubseteq \top \in \mathcal{O}$ **then** $S(C) \leftarrow S(C) \cup \{\top : \{(\emptyset, \{C \sqsubseteq \top\})\}\}$;

7     $\lfloor$ **else** $S(C) \leftarrow S(C) \cup \{\top : \{(\emptyset, \emptyset)\}\}$;

    // propagation

8 **repeat**

9   changed $\leftarrow$ **false**;

10   **foreach** $ax \in \mathcal{T} \cup \mathcal{A}$ **do**

11   **switch** typeof(ax) **do**

12   **case** *(NF1)* changed $\leftarrow$ changed | applyCR1(ax);

13   **case** *(NF2)* changed $\leftarrow$ changed | applyCR2(ax);

14   **case** *(NF3)* changed $\leftarrow$ changed | applyCR3(ax);

15   **case** *(NF4)* changed $\leftarrow$ changed | applyCR4(ax);

16   **case** *(NF5)* changed $\leftarrow$ changed | applyCR5(ax);

17   **case** *(NF6)* changed $\leftarrow$ changed | applyCR6(ax);

18 **until** changed $=$ **false**;

    // solution gathering

19 Sol $\leftarrow \{(\emptyset, \emptyset)\}$;

20 **foreach** $C1 \sqsubseteq D \in \mathcal{O}$ **do**

21     $\lfloor$ **if** $S(C1) \ni D : L$ **then** Sol $\leftarrow$ join(Sol,L);

22 **foreach** $C1 \sqsubseteq \exists r1.C2 \in \mathcal{O}$ **do**

23     $\lfloor$ **if** $R(r1) \ni (C1, C2) : L$ **then** Sol $\leftarrow$ join(Sol,L);

24 **return** meet-closure(Sol);

Procedure 3.4: `applyCR1`

Input: $ax = C1 \sqsubseteq D - a$ *(**NF1**)-axiom*
Output: **true** if the rule application was productive, **false** otherwise

1 changed $\leftarrow$ **false**;
2 **foreach** $C \in N_C^\top$ **do**
3   **if** $S(C) \ni C1 : L1$ **then**
4   **if** $S(C) \ni D :$ Lold **then** $L \leftarrow$ Lold;
5   **else** $L \leftarrow \{(\emptyset, \emptyset)\}$;
6   Lnew $\leftarrow$ join(L, meet(L1, $\{(\emptyset, \emptyset)\}$, ax, $C \sqsubseteq D$));
7   **if** Lnew $\neq$ L **then**
8              $S(C) \leftarrow S(C) \setminus \{D : L\} \cup \{D : Lnew\}$;
9   changed $\leftarrow$ **true**;

10 **return** changed;

Procedure 3.5: `applyCR2`

Input: $ax = C1 \sqcap C2 \sqsubseteq D - a$ *(**NF2**)-axiom*
Output: **true** if the rule application was productive, **false** otherwise

1 **foreach** $C \in N_C^\top$ **do**
2   **if** $S(C) \ni C1 : L1 \wedge S(C) \ni C2 : L2$ **then**
3   **if** $S(C) \ni D :$ Lold **then** $L \leftarrow$ Lold;
4   **else** $L \leftarrow \{(\emptyset, \emptyset)\}$;
5   Lnew $\leftarrow$ join(L, meet(L1, L2, ax, $C \sqsubseteq D$));
6   **if** Lnew $\neq$ L **then**
7              $S(C) \leftarrow S(C) \setminus \{D : L\} \cup \{D : Lnew\}$;
8   changed $\leftarrow$ **true**;

9 **return** changed;

Procedure 3.6: applyCR3

Input: ax = $C1 \sqsubseteq \exists r1.C2 - a$ **(NF3)**-*axiom*
Output: **true** if the rule application was productive, **false** otherwise

1 changed $\leftarrow$ **false**;
2 **foreach** $C \in N_C^\top$ **do**
3   **if** $S(C) \ni C1 : L1$ **then**
4   **if** $R(r1) \ni (C, C2) : Lold$ **then** $L \leftarrow Lold$;
5   **else** $L \leftarrow \{(\emptyset, \emptyset)\}$;
6   Lnew $\leftarrow$ join(L, meet(L1, $\{(\emptyset, \emptyset)\}$, ax, $C \sqsubseteq \exists r1.C2$));
7   **if** Lnew $\neq$ L **then**
8              $R(r1) \leftarrow R(r1) \setminus \{(C, C2) : L\} \cup \{(C, C2) : Lnew\}$;
9   changed $\leftarrow$ **true**;

10 **return** changed;

Procedure 3.7: applyCR4

Input: ax = $\exists r1.C2 \sqsubseteq D - a$ **(NF4)**-*axiom*
Output: **true** if the rule application was productive, **false** otherwise

1 **foreach** $C1 \in N_C^\top$ **do**
2   **if** $S(C1) \ni C2 : L1$ **then**
3   **foreach** $C \in N_C^\top$ **do**
4   **if** $R(r1) \ni (C, C1) : L2$ **then**
5   **if** $S(C) \ni D : Lold$ **then** $L \leftarrow Lold$;
6   **else** $L \leftarrow \{(\emptyset, \emptyset)\}$;
7   Lnew $\leftarrow$ join(L, meet(L1, L2, ax, $C \sqsubseteq D$));
8   **if** Lnew $\neq$ L **then**
9                      $S(C) \leftarrow S(C) \setminus \{D : L\} \cup \{D : Lnew\}$;
10   changed $\leftarrow$ **true**;

11 **return** changed;

Procedure 3.8: applyCR5

Input: ax $= r1 \sqsubseteq s - a$ *(NF5)-axiom*
Output: **true** if the rule application was productive, **false** otherwise

1 changed $\leftarrow$ **false**;
2 **foreach** $C \in N_C^\top$ **do**
3   **foreach** $D \in N_C^\top$ **do**
4   **if** $R(r1) \ni (C, D) : L1$ **then**
5   **if** $R(s) \ni (C, D) : $ Lold **then** $L \leftarrow $ Lold;
6   **else** $L \leftarrow \{(\emptyset, \emptyset)\}$;
7   Lnew $\leftarrow$ join(L, meet(L1, $\{(\emptyset, \emptyset)\}$, ax, $C \sqsubseteq \exists s.D$));
8   **if** Lnew $\neq$ L **then**
9              $R(s) \leftarrow R(s) \setminus \{(C, D) : L\} \cup \{(C, D) : $ Lnew$\}$;
10   changed $\leftarrow$ **true**;

11 **return** changed;

Procedure 3.9: applyCR6

Input: ax $= r1 \circ r2 \sqsubseteq s - a$ *(NF6)-axiom*
Output: **true** if the rule application was productive, **false** otherwise

1 **foreach** $C \in N_C^\top$ **do**
2   **foreach** $C1 \in N_C^\top$ **do**
3   **if** $R(r1) \ni (C, C1) : L1$ **then**
4   **foreach** $D \in N_C^\top$ **do**
5   **if** $R(r2) \ni (C1, D) : L2$ **then**
6   **if** $R(s) \ni (C, D) : $ Lold **then** $L \leftarrow $ Lold;
7   **else** $L \leftarrow \{(\emptyset, \emptyset)\}$;
8   Lnew $\leftarrow$ join(L, meet(L1, L2, ax, $C \sqsubseteq \exists s.D$));
9   **if** Lnew $\neq$ L **then**
10                   $R(s) \leftarrow R(s) \setminus \{(C, D) : L\} \cup \{(C, D) : $ Lnew$\}$;
11   changed $\leftarrow$ **true**;

12 **return** changed;

Function 3.10: join

Input: L1, L2 – *two label sets*
Output: L1 ⊕ L2

1 L ← L1 ∪ L2;
2 **return** L;

Function 3.11: meet

Input: L1, L2 – *two label sets*
    just, concl – *two normal form axioms*
Output: $L1 \otimes L2 \otimes \{(\{just\}^{\in \mathcal{A}?}, \{concl\}^{\in \mathcal{O}?})\}$

1 $L \leftarrow \{(A1 \cup A2, O1 \cup O2) \mid (A1, O1) \in L1 \wedge (A2, O2) \in L2\}$;
2 **if** (just ≠ **null** & just ∈ $\mathcal{A}$) **then**
3 ⌊ $L \leftarrow \{(A \cup \{just\}, O) \mid (A, O) \in L\}$;
4 **if** (concl ≠ **null** & concl ∈ $\mathcal{O}$) **then**
5 ⌊ $L \leftarrow \{(A, O \cup \{concl\}) \mid (A, O) \in L\}$;
6 **return** L;

Function 3.12: meet-closure

Input: L – *a label set*
Output: The closure of L under ⊗

1 **repeat**
2 │   size ← |L|;
3 │   L ← meet(L, L, **null**, **null**);
4 **until** size = |L|;
5 **return** remove-dominated(L);

Function 3.13: remove-dominated

Input: L – *a label set*
Output: The label set L with all ≺-dominated elements removed

1 **foreach** (A1, O1) ∈ L **do**
2 │   **foreach** (A2, O2) ∈ L **do**
3 │   ⌊ **if** (A1, O1) ≺ (A2, O2) **then** L ← L \ {(A2, O2)};
4 **return** L;

The solution set of a normalised relaxed abduction problem (in the sense of Proposition 3.12) expressed using the description logic $\mathcal{EL}^+$ can therefore be determined by running an extended completion-based classification procedure that a) permits abducible as well as told axioms to be used as justification for the completion rules, and b) extends the entries in the maps S and R with labels that encode solutions to the original problem. Next, we investigate the complexity of this approach.

Complexity Analysis

For the analysis of the complexity of algorithm RAPsolve-EL+, note that dominated label entries are not removed by meet or join. In this respect, our implementation deviates from the classic formulation of label correcting search for bi-criterion optimal (hyper)paths, where labels are minimized on every update (Guerriero & Musmanno, 2001). The effects of introducing such a pruning mechanism into our algorithm are analysed in detail in Section 3.3.3. In this line, we analyse the complexity of the basic algorithm without pruning.

Proposition 3.16 (Complexity of RAPsolve-EL+)
Let $\mathbf{RAP} = (\mathcal{T}, \mathcal{A}, \mathcal{O}, \preceq_\mathcal{A}, \preceq_\mathcal{O})$ *be an* $\mathcal{EL}^+$*-$\mathbf{RAP}$ and* $T_{\prec_\mathcal{A}}(n)$ *($T_{\prec_\mathcal{O}}(n)$) upper bounds on the complexity of comparing two sets of maximum size* n *w. r. t.* $\prec_\mathcal{A}$ *($\prec_\mathcal{O}$). Then* $T_{RAPsolve-EL+}(\mathbf{RAP})$, *the runtime for solving* $\mathbf{RAP}$ *using Algorithm 3.3, is*

$$O\left(2^{|\mathcal{A}|+|\mathcal{O}|} \cdot |N_\mathrm{C}|^2 \cdot |N_\mathrm{R}| + 2^{2 \cdot (|\mathcal{A}|+|\mathcal{O}|)} \cdot (T_{\prec_\mathcal{A}}(|\mathcal{A}|) + T_{\prec_\mathcal{O}}(|\mathcal{O}|))\right).$$

Proof. Note first that the sizes of the maps S and R are bounded by $2 \cdot |N_\mathrm{C}| \leq |S| \leq |N_\mathrm{C}| \cdot (|N_\mathrm{C}^\top|)$ and $|R| \leq |N_\mathrm{R}| \cdot |N_\mathrm{C}^\top|^2$. Each entry in S and R has a label whose maximum size may be $|L| \leq 2^{|\mathcal{A}|+|\mathcal{O}|}$ as no dominance is employed. The algorithm terminates at the latest when all labels have been generated, giving a bound of $|L| \cdot (|S| + |R|) \in O\left(2^{|\mathcal{A}|+|\mathcal{O}|} \cdot |N_\mathrm{C}|^2 \cdot |N_\mathrm{R}|\right)$ operations for the completion. Closing the intermediate result *Sol* under $\otimes$ is quadratic in $|Sol|$ and can be bounded by $O\left(2^{2 \cdot (|\mathcal{A}|+|\mathcal{O}|)}\right)$; the final minimization requires a number $\prec$-comparisons that is quadratic in $|Sol|$ again. □

While there is potential for a reduction in complexity by pruning intermediate results (see the next section), it is unlikely that an efficient algorithm for this problem exists: As shown in Skriver (2000), the bi-criterion shortest path problem is NP-hard, which obviously carries over to the hyperpath variant as the number of potential solutions grows. For relating this result to the one derived for the generic approach (Proposition 3.11), assume that all axioms are normalised $\mathcal{EL}^+$ axioms. It is then easy to show that $|\mathcal{T}|, |\mathcal{A}| \in O\left(|N_C|^3 + |N_R|^3\right)$ and $|\mathcal{O}| \in O\left(|N_C|^2 \cdot |N_R|\right)$. Moreover T_L^{class} is polynomial. Solution generation for any of the approaches then has a complexity of $O\left(2^{|N_C|^3 + |N_R|^3} \cdot poly(|N_C|, |N_R|)\right)$. To get a clearer picture of the true complexities, a comparative evaluation of both approaches is provided in Section 4.2. in a nutshell, the main difference on the conceptual level is that the generic approach uses an exponential number of normal classification operations, whereas the specific $\mathcal{EL}^+$ solution requires just a single classification run over a (potentially exponentially) labelled structure that encodes all alternatives. Moreover, the black-box approach basically performs an uninformed search in the space spanned by $\mathcal{A}$; the glass-box approach instead makes use of the domain information represented by $\mathcal{T}$ and $\mathcal{A}$, excluding parts of the hypergraph not reachable from $V_\mathcal{T}$ from the search.

To conclude the discussion on the computational properties of the basic algorithm, it should be pointed out that the complexity of finding $\prec$-dominated tuples may vary widely: While the cardinality or the sum-of-weights of a set can be determined virtually "for free" using an auxiliary variable that is updated on adding elements, inclusion of two sets can typically be determined in linear time if the sets are ordered. Testing entailment of two sets, however, is typically more complicated, requiring for example polynomial time for $\mathcal{EL}^+$. Although optimisations are possible, for example, by utilizing a reasoner that implements incremental classification, there remains a considerable effort to be taken into account since up to $\left(2^{|\mathcal{A}| + |\mathcal{O}|}\right)^2$ such tests may be necessary.

3.3.3 Employing Pruning to Increase Efficiency

As mentioned already, most formulations of label-correcting algorithms for multi-criterion shortest path search assume that node labels are minimal at all times; this is typically realized by removing dominated entries after each label update (c.f. Guerriero & Musmanno, 2001). As it is the case for any heuristic function used for pruning, it is essential to ensure that completeness is retained in this process. In this section we analyse the effects of extending RAPsolve-EL+ with pruning, dependent on the choice of $\preceq_\mathcal{A}$ and $\preceq_\mathcal{O}$.[11]

---

[11] For the black-box variant, RAPsolve-Generic, methods for pruning unnecessary classifications and entailment tests has been presented in Section 3.3.1.

For integrating pruning into the algorithm, observe that `join` is the final operation in updating labels, and that it is always used to extend an existing label (first parameter) with new information (second parameter). If we employ pruning, we may therefore assume that the first label is domination-free, and only need to compare the new label entries with the old ones. This intuition is implemented by `join-and-prune` (Algorithm 3.14). To extend Algorithm 3.3 with pruning, everything that needs to be done is to replace all calls to `join` by calls to `join-and-prune`.

Function 3.14: `join-and-prune`

Input: L1, L2 – *two label sets, the first is assumed dominance-free*
Output: $L1 \oplus L2$ – *with all $\prec$-dominated elements removed*

1 $L \leftarrow L1$;
2 **foreach** $(A2, O2) \in L2$ **do**
3 $\quad$ Ldom $\leftarrow \emptyset$;
4 $\quad$ **foreach** $(A, O) \in L$ **do**
5 $\quad\quad$ **if** $(A, O) \prec (A2, O2)$ **then** Ldom $\leftarrow$ Ldom $\cup \{(A2, O2)\}$;
6 $\quad\quad$ **else if** $(A2, O2) \prec (A, O)$ **then** Ldom $\leftarrow$ Ldom $\cup \{(A, O)\}$;
7 $\quad$ $L \leftarrow (L \cup \{(A2, O2)\}) \setminus$ Ldom;
8 **return** L;

Definition 3.9 (Safety of $\preceq$ for pruning)
Let $\mathbf{RAP} = (\mathcal{T}, \mathcal{A}, \mathcal{O}, \preceq_\mathcal{A}, \preceq_\mathcal{O})$ *be a relaxed abduction problem and $\preceq$ the preorder induced by $\preceq_\mathcal{A}$ and $\preceq_\mathcal{O}$ according to Definition 3.1. We call $\preceq$ safe for pruning if and only if* $(A_1, O_1) \preceq (A_2, O_2)$ *implies* $(A_1 \cup A^*, O_1 \cup O^*) \preceq (A_2 \cup A^*, O_2 \cup O^*)$ *for all pairs* $(A_1, O_1), (A_2, O_2), (A^*, O^*)$.

Definition 3.9 guarantees that no tuple that was dropped due to dominance could have been extended to a dominating solution by extending the hyperpath (using the $\otimes$ operation). Based on this definition, we now revisit the different instantiations of **RAP** presented in Section 3.2.4.

Proposition 3.17 (Pruning-safety of selected preorders)
Let $\mathbf{RAP} = (\mathcal{T}, \mathcal{A}, \mathcal{O}, \preceq_\mathcal{A}, \preceq_\mathcal{O})$ *be an* $\mathcal{EL}^+$ *relaxed abduction problem. The following properties hold:*
> *a) The preorder $\preceq^s$ induced by $(\mathcal{T}, \mathcal{A}, \mathcal{O}, \subseteq, \supseteq)$ is safe for pruning.*
> *b) The preorder $\preceq^e$ induced by $(\mathcal{T}, \mathcal{A}, \mathcal{O}, \dashv, \models)$ is safe for pruning.*

c) The preorder $\preceq^{e^*}$ induced by $(\mathcal{T}, \mathcal{A}, \mathcal{O}, \models_{\mathcal{T}}, \models_{\mathcal{T}})$ is safe for pruning.

d) The preorder $\preceq^c$ induced by $(\mathcal{T}, \mathcal{A}, \mathcal{O}, \leq^{|\cdot|}, \geq^{|\cdot|})$ is not safe for pruning.

e) The preorder $\preceq^w$ induced by $(\mathcal{T}, \mathcal{A}, \mathcal{O}, \leq^{\varsigma_A}, \geq^{\varsigma_O})$ is not safe for pruning (for arbitrary weight systems).

Proof. To show claim a) let $(A_1, O_1) \preceq^s (A_2, O_2)$. This implies $A_1 \subseteq A_2 \wedge O_1 \supseteq O_2$. By monotonicity of set inclusion then also $A_1 \cup A^* \subseteq A_2 \cup A^* \wedge O_1 \cup O^* \supseteq O_2 \cup O^*$, giving $(A_1 \cup A^*, O_1 \cup O^*) \preceq^s (A_2 \cup A^*, O_2 \cup O^*)$. For claim b) assume $(A_1, O_1) \preceq^e (A_2, O_2)$, which implies $A_1 \models A_2 \wedge O_1 \models O_2$. Then also $A_1 \cup A^* \models A_2 \cup A^* \wedge O_1 \cup O^* \models O_2 \cup O^*$, which is by definition equivalent to $(A_1 \cup A^*, O_1 \cup O^*) \preceq^e (A_2 \cup A^*, O_2 \cup O^*)$. Property c) is analogous.

We show d) by a counterexample: Let $(A_1, O_1) = (\{a_1\}, \{o_1\})$, $(A_2, O_2) = (\{a_2, a_3\}, \{o_1\})$, and $(A^*, O^*) = (\{a_2, a_3\}, \{o_2\})\}$. Then $(A_1, O_1) \prec^c (A_2, O_2)$, but $(A_2 \cup A^*, O_2 \cup O^*)^c (A_1 \cup A^*, O_1 \cup O^*)$.

Since $(\mathcal{T}, \mathcal{A}, \mathcal{O}, \leq^{|\cdot|}, \geq^{|\cdot|})$ of $(\mathcal{T}, \mathcal{A}, \mathcal{O}, \leq^{\varsigma_A}, \geq^{\varsigma_O})$ with identical weight 1 for all axioms and combination by addition, e) is implied by d). However, this choice is not the only problematic one, a structurally similar counterexample can easily be constructed, e.g. for Zadeh style fuzzy semantics. □

This shows that the subsumption-based and the entailment-based instantiations lead to a notion of dominance that can be used for pruning without sacrificing completeness, while this does not hold for the other examples considered in Section 3.2.4. Note, however, that Definition 3.9 is defined using $\preceq$ and therefore only guarantees that a dominated tuple $(A_1, O_1) \prec (A_2, O_2)$ can never be part of a solution $(A_1 \cup A^*, O_1 \cup O^*)$ that is *better* than the analogous extension $(A_2 \cup A^*, O_2 \cup O^*)$ of the dominating tuple. Yet, the situation $(A_1 \cup A^*, O_1 \cup O^*) \simeq (A_2 \cup A^*, O_2 \cup O^*)$ is not excluded, implying that $\preceq$-equal tuples may be lost due to pruning. It should be pointed out that this occurs for all instantiations of **RAP** including the most general one, as can be seen by letting $A^* \supseteq A_2 \setminus A_1$ and $O^* \supseteq O_1 \setminus O_2$; a stronger definition of safety based on $\prec$ would therefore not be met by any of the instantiations. The effects of this observation heavily depend on the choice of preorders, as summed up in Proposition 3.18.

To demonstrate the effects of different pruning methods practically, we now return to the example diagnostics scenario:

Example 3.5 (Effects of Pruning)
Assume that in the context of our running diagnostic example, we are given

an error hierarchy expressed using abducible axioms[12]:

$$\text{PowerSupplyFluctuations} \sqsubseteq \text{PowerSystemError}$$
$$\text{PowerSystemError} \sqsubseteq \text{CriticalError}$$
$$\text{PowerSupplyFluctuations} \sqsubseteq \text{CriticalError}$$

In the derivation process the pair (PowerSupplyFluctuations, CriticalError) *will be derived twice using completion rule (CR1), namely once using the first two axioms, and once using the third alone. The label of the S-entry will then be*

$$L(\text{PowerSupplyFluctuations}, \text{CriticalError}) =$$
$$\{\{\text{PowerSupplyFluctuations} \sqsubseteq \text{PowerSystemError},$$
$$\text{PowerSystemError} \sqsubseteq \text{CriticalError}\},$$
$$\{\text{PowerSupplyFluctuations} \sqsubseteq \text{CriticalError}\}\}$$

If either no pruning or subset-based pruning is employed in the algorithm, this label set is not affected by pruning and propagated in the graph as is. However, in the case of entailment-based pruning, the label will in fact be reduced to:

$$L(\text{PowerSupplyFluctuations}, \text{CriticalError}) =$$
$$\{\{\text{PowerSupplyFluctuations} \sqsubseteq \text{CriticalError}\}\}$$

This cuts down the size of this label (and, thus, also the effort required for derivation step building on this S-entry) significantly.

Next, we investigate the effects of subset-based and entailment-based pruning methods in more detail:

Proposition 3.18 (Effects of pruning)
Let **RAP** $= (\mathcal{T}, \mathcal{A}, \mathcal{O}, \preceq_{\mathcal{A}}, \preceq_{\mathcal{O}})$ *be an* $\mathcal{EL}^+$ *relaxed abduction problem. The following properties hold:*

a) *Pruning can used for solving* $(\mathcal{T}, \mathcal{A}, \mathcal{O}, \subseteq, \supseteq)$ *without changing* $Sol_{\mathbf{RAP}}$.

b) *If pruning is employed in solving* $(\mathcal{T}, \mathcal{A}, \mathcal{O}, \sqsupseteq, \models)$, *the resulting solution set is semantically equivalent to the one found without pruning, but not necessarily identical.*

c) *If pruning is employed in solving* $(\mathcal{T}, \mathcal{A}, \mathcal{O}, \sqsupseteq_{\mathcal{T}}, \models_{\mathcal{T}})$, *the resulting solution set is semantically equivalent to the one found without pruning, but not necessarily identical.*

---

[12]To keep this example as simple as possible, we intentionally do not consider effects with the other, existing axioms in this example.

Proof Property a) follows from the fact that at most $\simeq^s$-equivalent tuples can be lost due to Proposition 3.17 a), and $(A, O) \simeq^s (A', O') \leftrightarrow A = A' \wedge O = O'$. For claim b), note that only $\simeq^e$-equivalent tuples can be lost due to Proposition 3.17 b). Here, $(A, O) \simeq^s (A', O')$ if and only if $A \equiv A' \wedge O \equiv O'$. This means that different syntactical representations of a semantically equivalent solution (i. e. different derivations) may be lost due to dominance. As an example consider the situation $\mathcal{T} = \emptyset, A_1 = \{C \sqsubseteq X, X \sqsubseteq F\}, A_2 = \{C \sqsubseteq E\}, O = \{C \sqsubseteq E\}$. Obviously $(A_2, O) \prec (A_1, O)$, so the latter would be pruned. Now let $A^* = \{E \sqsubseteq X, X \sqsubseteq C\}, O^* = \{E \sqsubseteq C\}$. Then $A_1 \cup A^* \equiv A_2 \cup A^* \equiv \{C \equiv E, C \equiv X, E \equiv X\}$ and $O_1 \cup O^* \equiv O_2 \cup O^* \equiv \{C \equiv E\}$, the extensions are therefore equivalent but not syntactically identical. The same is true for c). $\qquad\square$

Despite these results, it is possible to extend arbitrary instantiations **RAP** with pruning without losing correctness: Remembering that $\preceq_\mathcal{A}$ and $\preceq_\mathcal{O}$ are required to be monotone respectively anti-monotone for set inclusion, it is obvious that pruning w. r. t. set inclusion may always be applied safely, independent of the concrete preorders chosen. Algorithm 3.15 presents an adapted variant of `join-and-prune` implementing this idea. We conclude our analysis of pruning in the context of relaxed abduction for $\mathcal{EL}^+$ knowledge bases with an analysis of the computational benefits gained by introducing safe pruning based on set inclusion in Proposition 3.19.

Function 3.15: `join-and-prune`'   (safe)

Input: L1, L2 – *two label sets, the first is assumed dominance-free*
Output: L1 $\oplus$ L2 – *with all inclusion-dominated elements removed*

```
1  L ← L1;
2  foreach (A2, O2) ∈ L2 do
3  |   Ldom ← ∅;
4  |   foreach (A, O) ∈ L do
5  |   |   if (A ⊆ A2 & O ⊃ O2) | (A ⊂ A2 & O ⊇ O2) then
6  |   |   |   Ldom ← Ldom ∪ {(A2, O2)};
7  |   |   else if (A ⊇ A2 & O ⊂ O2) | (A ⊃ A2 & O ⊆ O2) then
8  |   |   |   Ldom ← Ldom ∪ {(A, O)};
9  |   L ← (L ∪ {(A2, O2)}) \ Ldom;
10 return L;
```

Proposition 3.19 (Complexity of RAPsolve-ELPlus with safe pruning)

Let $\mathbf{RAP} = (\mathcal{T}, \mathcal{A}, \mathcal{O}, \preceq_\mathcal{A}, \preceq_\mathcal{O})$ *be an* $\mathcal{EL}^+$ *relaxed abduction problem and* $T_{\prec_\mathcal{A}}(n)$ *(*$T_{\prec_\mathcal{O}}(n)$*) upper bounds on the complexity of comparing two sets of maximum size* n *w. r. t.* $\prec_\mathcal{A}$ *(*$\prec_\mathcal{O}$*). Then* $T_{RAPsolve-EL+^{sp}}(\mathbf{RAP})$*, the runtime for solving* $\mathbf{RAP}$ *using Algorithm 3.3 extended with safe pruning, is*

$$O\left(2^{|\mathcal{A}|} \cdot |N_\mathrm{C}|^2 \cdot |N_\mathrm{R}| + 4^{|\mathcal{A}|} \cdot (T_{\prec_\mathcal{A}}(|\mathcal{A}|) + T_{\prec_\mathcal{O}}(|\mathcal{O}|))\right).$$

Moreover, the expected runtime[13] $E(T_{RAPsolve-EL+^{sp}}(\mathbf{RAP}))$*, is given by*

$$1.5^{|\mathcal{A}|} \cdot |N_\mathrm{C}|^2 \cdot |N_\mathrm{R}| + 2.25^{|\mathcal{A}|} \cdot (T_{\prec_\mathcal{A}}(|\mathcal{A}|) + T_{\prec_\mathcal{O}}(|\mathcal{O}|)).$$

Proof. As before (Proposition 3.16), it holds that $2 \cdot |N_\mathrm{C}| \leq |S| \leq |N_\mathrm{C}| \cdot (|N_\mathrm{C}^\top|)$ and $|R| \leq |N_\mathrm{R}| \cdot |N_\mathrm{C}^\top|^2$. Since label entries represent potential solutions, using pruning based on set inclusion reduces the maximum label size to $O\left(2^{|\mathcal{A}|} \cdot \min\left\{\frac{2^{|\mathcal{O}|}}{\sqrt{|\mathcal{A}|}}, 1\right\}\right)$ by Proposition 3.4 c). Therefore, the completion requires at most $|L| \cdot (|S| + |R|) \in O\left(2^{|\mathcal{A}|} \cdot \min\left\{\frac{2^{|\mathcal{O}|}}{\sqrt{|\mathcal{A}|}}, 1\right\} \cdot |N_\mathrm{C}|^2 \cdot |N_\mathrm{R}|\right)$ operations. Closing the intermediate result *Sol* under $\otimes$ and the final minimization using $\prec$ are both quadratic in $|Sol|$, giving an overall complexity of $O\left(\left(2^{|\mathcal{A}|} \cdot \min\left\{\frac{2^{|\mathcal{O}|}}{\sqrt{|\mathcal{A}|}}, 1\right\}\right)^2 \cdot (T_{\prec_\mathcal{A}}(|\mathcal{A}|) + T_{\prec_\mathcal{O}}(|\mathcal{O}|))\right)$. Since the min term will only be less than 1 for very atypical situations we drop it, giving the worst-case result.

For the proof of the expected runtime result, we determine the average size of a label that has been minimized w. r. t. set inclusion. As the bound for the O component is constant, we only need to consider the assumption set. Let $\preceq_\mathcal{A}$ be arbitrary, and represent every subset $A \subseteq \mathcal{A}$ as a binary vector of length $|\mathcal{A}|$. We now determine the expected number $A(n, l)$ of maximal $(0, 1)$ vectors of length $l = |\mathcal{A}|$ among a set of n distinct such vectors chosen uniformly at random. Following the approach outlined in Bentley et al. (1978), $A(n, l)$ can be expressed by the recurrence $A(n, l) \leq \lceil\frac{n}{2}\rceil \cdot \frac{A(n, l-1)}{n} + \lceil\frac{n}{2}\rceil \cdot \frac{A(\lceil n/2\rceil, l-1)}{\lceil n/2\rceil} \approx \frac{1}{2} \cdot A(n, l-1) + A(n/2, l-1)$ which can be understood as follows: Let the vectors be arranged in a $(n \times l)$ matrix sorted by the first component. A randomly chosen vector $\vec{v}$ starts with 1 or 0 with probability 0.5 each.

---

[13]Note that we do not consider any specific distribution over the inputs. Instead, we use an estimation of maximum label size given completely random input.

In the former case, $\vec{v}$ cannot be dominated by any vector starting with a 0, which instantly rules the "lower half" of the table out, and fixes its probability of being dominated by another vector starting with 1 to given by the expected number of maxima among the remaining $\lceil n/2 \rceil$ vectors divided by their number; taken together $\vec{v}$ is maximal with probability $A(\lceil n/2 \rceil, l-1)/\lceil n/2 \rceil$. If $\vec{v}$ starts with 0 instead, there can be no other vectors ruled out based on the first component only; we can thus determine its probability of being maximal by $A(n, l-1)/n$. Summing up these probabilities and multiplying the result by the number n of original vectors yields the recurrence $A(n,l) \lesssim \frac{1}{2} \cdot A(n, l-1) + A(n/2, l-1)$ given above; its base values being $A(k,1) = A(1,k) = 1$. For the worst-case analysis we assume $n \geq 2^{l-1}$ (running into the "long tail" of the recursion), the recursion then terminates when $l = 1$ at recursion depth $l-1$. An upper bound is thus given by $A(n,l) \leq A(\max\{n,1\}, l-1) + \frac{1}{2} \cdot A(\max\{n,1\}, l-1) = \frac{3}{2} \cdot A(\max\{n,1\}, l-1) = (\frac{3}{2})^{l-1}$. Taken together, this leads to an expected label size of $O(1.5^{|\mathcal{A}|})$. □

To wrap up the results in this section: Pruning can be introduced without sacrificing correctness for the inclusion- and entailment-based instantiations. In case of the former, the solution set determined with pruning is identical to the one determined without pruning, whereas pruning based on entailment may drop syntactically different representations of semantically equivalent solutions. Since all preorders must be compatible with set inclusion by definition, we propose to always use pruning based on set inclusion during construction, and only minimize the final result set w. r. t. the user-provided preferences $\preceq_{\mathcal{A}}$ and $\preceq_{\mathcal{O}}$. Although relaxed abduction is inherently an ExpTime problem, this approach provides a solution with relatively small exponentials ($2^{|\mathcal{A}|}$ for creation, $4^{|\mathcal{A}|}$ for the final minimization). As our average-case analysis shows, expected runtime is still significantly better (exponentials $1.5^{|\mathcal{A}|}$ for creation, $2.25^{|\mathcal{A}|}$ for the final minimization). Whether these relatively small exponentials do indeed "make nice pets" and lead to a solution which can be used for practical applications is evaluated in Section 4.2.

3.3.4 Axiom Selection Strategies

Algorithm 3.3 iterates over all axioms in $\mathcal{T} \cup \mathcal{A}$ in line 10, without any specific order. However, if we use pruning for keeping the labels minimized at all time, the *axiom selection strategy* that determines which axiom is chosen for propagation next may have a significant effect on the typical label size, and therefore on the computational cost of `meet` and `join-and-prune` operations. A look at the completion rules in Figure 3.6 and the definition of `meet` (Algorithm 3.11) shows that it is always better to explore $\mathcal{T}$-propagations (edges in $E^{\mathcal{T}}$) before $\mathcal{A}$-propagations (edges in $E^{\mathcal{A}}$), since the former are "for free" in the sense that their weight never adds to the first component of a label entry. By generating partial solutions requiring few assumptions first, more expensive derivations of the same observation set can be pruned early before they are propagated through the hypergraph, which would require a larger number of comparisons later on. We shortly discuss three strategies for axiom selection here.

Axiom sorting. The arguably most straightforward method for favouring $E^{\mathcal{T}}$-edges is to order the axioms in $\mathcal{T} \cup \mathcal{A}$ in such a way that axioms from $\mathcal{T}$ precede axioms from $\mathcal{A}$. This ensures that single derivation steps are always tried assumption-less first, introducing abducible axioms only if required on a step-by-step basis. It can for example help prevent the exploration of $E^{\mathcal{A}}$-edges in situations where an axiom is both in $\mathcal{T}$ and $\mathcal{A}$, or where $\mathcal{A}$ contains a variant of a $\mathcal{T}$-axiom with weaker premise. However, this solution works on a per-edge basis: In situations where, for instance, a single $E^{\mathcal{A}}$-edge could be replaced by a path consisting of several $E^{\mathcal{T}}$-edges, the "expensive" one-step solution will be explored first, and be replaced by the assumption-less path consecutively.

$\mathcal{T}$-materialisation. Another easy to implement strategy that improves on this point consists of duplicating the propagation loop in 3.3, and have the first copy iterate over the elements of knowledge base $\mathcal{T}$ only. This way, everything that can be derived without using any assumptions at all is materialized in the sets S and R before assumptions are introduced (all label entries have $\emptyset$ as first component at that point). The second copy of the propagation loop then picks axioms from $\mathcal{T} \cup \mathcal{A}$ as before, integrating the additional derivations made possible by introducing assumptions. To retain completeness, it is vital that the second loop iterates over both knowledge bases, otherwise derivations might get lost (if a $\mathcal{A}$-justified derivation introduces a new label entry, this entry can be a source for new $\mathcal{T}$-derivations). In this second loop, elements may be picked following the axiom sorting strategy. This strategy provides a good tradeoff between effectiveness of pruning and ease of implementation.

$\mathcal{A}$-interleaving. In an extension of the materialisation strategy, one could start by completely propagating $\mathcal{T}$ until saturation is reached, and repeat this materialisation step after each productive propagation of some $\mathcal{A}$-axiom, where productive means that a new entry was introduced to S or R, or the label of an existing entry was updated. This way, the information contained in an assumption is completely exploited before another assumption is allowed; derivation steps requiring assumptions are only interleaved if no other propagations can be done. We conjecture that this approach is optimal in the sense that without having any further information on the axioms, no other generic axiom selection strategy introduces less assumptions that are later dropped due to pruning. However, $\mathcal{A}$-interleaving is more complex to implement than the other approaches, as the skeleton below (which replaces the propagation loop in Algorithm 3.3) shows. Moreover, it is not clear whether the increased number of rule applicability checks is weighed off by the reduced label size in practice.

```
// propagation
repeat
    repeat
        changed ← false;
        foreach ax ∈ 𝒯 do
            // call applyCR* for ax, set changed to result
    until changed = false;
    foreach ax ∈ 𝒜 do
        // call applyCR* for ax, set changed to result
        if changed = true then break;
until changed = false;
```

We conclude Section 3.3 by revisiting the implementation-related requirements defined in Section 1.2: Requirement **R6** is met by both algorithms provided, as we have shown in Propositions 3.10 and 3.15. We also answer the question of complexity (Requirement **R7**) positively for Algorithm 3.3: Although the analysis in Proposition 3.16 shows an exponential behaviour, these exponentials were shown to be small – a characteristic shared by many other algorithms employed practically in computer science. Regarding the generic variant in Algorithm 3.1, the answer to whether Requirement **R7** is met largely depends on the concrete description logic chosen due to the exponential number of classification operations required. However, the choice of DL can be adapted in a very fine-grained way to the problem at hand.

3.4 Extensions to the Basic Setting

Having introduced the basic framework of relaxed abduction and two algorithms for solving relaxed abduction problems, we now turn towards several extensions to the basic setting which are relevant to our envisioned applications.

3.4.1 Extending Completion-based Relaxed Abduction to More Expressive DLs

The generic solution admits the use of arbitrary description logics, the completion-based method implemented in Algorithm 3.3 is currently limited to the description logic $\mathcal{EL}^+$. Although the expressive means provided by this DL have been shown to be sufficient for modelling large taxonomic structures such as the SNOMED CT and GALEN medical ontologies, advanced applications may incur the need for additional features of a domain to be represented. In this section, we study how the completion-based approach can be extended accordingly.

Relaxed Abduction over $\mathcal{EL}^{++}$ Knowledge Bases. The increase in expressiveness from $\mathcal{EL}^+$ to $\mathcal{EL}^{++}$ is due to the existence of the $\bot$ concept that allows to state the disjointness of concepts, and the availability of nominals $\{a\}$ which make it possible to express information about individuals, i.e. to formulate assertional in addition to taxonomical information (see Baader et al., 2005a). To keep the presentation simple we do not include concrete domains in our discussion, they can however be added to the formal framework easily. Table 3.2 sums up the constructors available in $\mathcal{EL}^{++}$, with new features marked by a leading asterisk. As before, A signifies an atomic concept, C, D represent arbitrary concepts, and r_i, r, s are roles; the new name a represents an individual taken from the set N_{I} of individual names. Note that due to the availability of nominals and concept disjunction, all $\mathcal{EL}^{++}$ standard reasoning tasks can be reduced to subsumption checking (c.f. Baader et al., 2005b). Therefore, relaxed abduction over $\mathcal{EL}^{++}$ knowledge bases can be used to determine explanations for all types of axioms including concept membership assertions $C(a)$ (which can be rewritten as $\{a\} \sqsubseteq C$)

Table 3.2: $\mathcal{EL}^{++}$ syntax and semantics

	Syntax	Semantics
top concept	$\top$	$\Delta^{\mathcal{I}}$
(*) bottom concept	$\bot$	$\emptyset$
(*) nominal	$\{a\}$	$\{a^{\mathcal{I}}\}$
atomic concept	A	$A^{\mathcal{I}}$
concept conjunction	$C \sqcap D$	$C^{\mathcal{I}} \cap D^{\mathcal{I}}$
existential restriction	$\exists r . C$	$\{x \in \Delta^{\mathcal{I}} \mid \exists y \in \Delta^{\mathcal{I}} :$
		$(x,y) \in r^{\mathcal{I}} \wedge y \in C^{\mathcal{I}}\}$
concept inclusion axiom	$C \sqsubseteq D$	$C^{\mathcal{I}} \subseteq D^{\mathcal{I}}$
role inclusion axiom	$r_1 \circ \cdots \circ r_n \sqsubseteq s$	$r_1^{\mathcal{I}} \circ \cdots \circ r_n^{\mathcal{I}} \subseteq s^{\mathcal{I}}$

and role membership assertions.[14] The *unique name assumption* frequently employed in the context of description logics (but nor necessarily in OWL) says that two distinct individuals must refer to distinct domain elements. This can be enforced by extending $\mathcal{T}$ with axioms $\{a\} \sqcap \{b\} \sqsubseteq \bot$ for all $a, b \in N_I$ s.t. $a \neq b$.

Similar to $\mathcal{EL}^+$, any $\mathcal{EL}^{++}$ knowledge base can be transformed in linear time into an equivalent one that only contains axioms of the form **(NF1)** $C_1 \sqsubseteq D$, **(NF2)** $C_1 \sqcap C_2 \sqsubseteq D$, **(NF3)** $C_1 \sqsubseteq \exists r_1 . C_2$, **(NF4)** $\exists r_1 . C_2 \sqsubseteq D$, **(NF5)** $r_1 \sqsubseteq s$, and **(NF6)** $r_1 \circ r_2 \sqsubseteq s$. For $\mathcal{EL}^{++}$, C_1, C_2 may be a concept name including $\top$ or a nominal; D may be any of the aforementioned, or $\bot$. The consequence-driven approach for solving relaxed abduction can be extended to $\mathcal{EL}^{++}$ based on the set of $\mathcal{EL}^{++}$ completion rules depicted in Figure 3.7. The notation $C_1 \rightsquigarrow_R C_2$ used in rule **(CR8)** hereby signifies that there is a chain of R-entries connecting C_1 (or an arbitrary nominal $\{b\}$) to C_2.

There is a number of considerations that have to be taken into account when extending the consequence-driven algorithm for relaxed abduction to this more expressive formalism. We discuss these considerations next, and investigate how Algorithm 3.3 can be extended accordingly. First, note that the availability of both nominals and the $\bot$ concept may render a knowledge base inconsistent if the KB entails a nominal to be subsumed by $\bot$. If $\mathcal{T}$ inconsistent (i.e. $\mathcal{T} \models \bot$, which can easily be determined before starting

---

[14]In fact, for most practical applications it suffices to only allow these types of axioms for $\mathcal{O}$, since all observational information on concrete instances can be expressed this way.

$$(\text{IR1}) \ \frac{}{C \sqsubseteq C} \qquad (\text{IR2}) \ \frac{}{C \sqsubseteq \top}$$

$$(\text{CR1}) \ \frac{C \sqsubseteq C_1}{C \sqsubseteq D} \ [C_1 \sqsubseteq D \in \mathcal{T}]$$

$$(\text{CR2}) \ \frac{C \sqsubseteq C_1 \qquad C \sqsubseteq C_2}{C \sqsubseteq D} \ [C_1 \sqcap C_2 \sqsubseteq D \in \mathcal{T}]$$

$$(\text{CR3}) \ \frac{C \sqsubseteq C_1}{C \sqsubseteq \exists r_1 . C_2} \ [C_1 \sqsubseteq \exists r_1 . C_2 \in \mathcal{T}]$$

$$(\text{CR4}) \ \frac{C \sqsubseteq \exists r_1 . C_1 \qquad C_1 \sqsubseteq C_2}{C \sqsubseteq D} \ [\exists r_1 . C_2 \sqsubseteq D \in \mathcal{T}]$$

$$(\text{CR5}) \ \frac{C \sqsubseteq \exists r_1 . D}{C \sqsubseteq \exists s . D} \ [r_1 \sqsubseteq s \in \mathcal{T}]$$

$$(\text{CR6}) \ \frac{C \sqsubseteq \exists r_1 . C_1 \qquad C_1 \sqsubseteq \exists r_2 . D}{C \sqsubseteq \exists s . D} \ [r_1 \circ r_2 \sqsubseteq s \in \mathcal{T}]$$

$$(\text{CR7}) \ \frac{C \sqsubseteq \exists r_1 . C_1 \qquad C_1 \sqsubseteq \bot}{C \sqsubseteq \bot}$$

$$(\text{CR8}) \ \frac{C_1 \sqsubseteq \{a\} \qquad C_2 \sqsubseteq \{a\} \qquad C_1 \rightsquigarrow_R C_2 \qquad C_2 \sqsubseteq D}{C_1 \sqsubseteq D}$$

Figure 3.7: Completion rules for classification in $\mathcal{EL}^{++}$

the completion-based algorithm) the $\mathcal{R}AP$ obviously has an empty solution set which can be returned instantly. Inconsistencies based on subsets of $\mathcal{A}$ are handled as follows:

1. To identify inconsistencies efficiently, the completion-based algorithm for solving $\mathcal{EL}^{++}$ relaxed abduction problems maintains a list No of *nogoods*, i.e. of inclusion-minimal sets $A \subseteq \mathcal{A}$ such that $\mathcal{T} \cup A$ is inconsistent. No is initially empty, and every time $\bot : (A, O)$ is added to $S(\{b\})$ for any nominal $\{b\}$, the set A is added to No. The idea of explicitly storing inconsistency information to allow for an early pruning of search branches has already been applied successfully in SAT solvers. Whereas SAT solvers do not necessarily store this information in minimised form, we adopted the ATMS idea of storing minimised labels for the nogood set to minimise space requirements.

2. All node labels are pruned w.r.t. No; meaning that any label entry (A, O) with $A \supseteq N$ for any $N \in No$ is removed. This can be realized eagerly by pruning all labels when a new nogood is detected (minimising label size at possibly high computational cost), or in a lazy approach where nogood-dominated labels are identified and dropped during propagation and solution gathering.

Moreover, since $\bot$ is by definition subsumed by every other concept, $C \sqsubseteq D$ is no longer only entailed if $S(C) \ni D$, but also if $S(C) \ni \bot$ (which is equivalent to $C \equiv \bot$). Solution gathering must take this into account.

3. To this end, the loop over **(NF1)**-observations during solution gathering (line 20) is extended to take unsatisfiable concepts into account. This is realised by additionally looking for $S(C_1) \ni \bot : L$, and in case such an entry exists adding the label L (extended by the observation $C \sqsubseteq D$) to *Sol*.

Finally, propagation of information on nominals as well as triggering of the two additional rules **(CR7)** and **(CR8)** must be guaranteed.

4. All loops iterating over concept names from $N_C^\top$ must be modified to iterate over concept names from $N_C^\top$ as well as nominals, and potentially the bottom concept (depending on the type of axiom). We denote the set of all nominals by $N_N = \{\{a\} \mid a \in N_I\}$.

5. Note that neither **(CR7)** nor **(CR8)** is guarded by an axiom, these rules must therefore be triggered explicitly, independently of the axiom loop.

Algorithm 3.16 presents our extended algorithm for solving $\mathcal{EL}^{++}$ relaxed abduction problems based on the lazy nogood handling approach, and a minimized storage of the nogoods (supersets of nogoods are omitted). New and updated procedures and functions are presented subsequently. To avoid repetition we omit unchanged lines of code where necessary (indicated by a comment); furthermore we do not provide pseudo code for the natural extension of meet to four label sets used in Algorithm 3.21. As can easily be shown, Propositions 3.15 and 3.18 carry over straightforwardly from RAPSolve-EL+ to RAPSolve-EL++, the only difference being that paths may be blocked when their weight corresponds to an inconsistent knowledge base w.r.t. $\mathcal{T}$. Similarly, the complexity results in Propositions 3.16 and 3.19 carry over to $\mathcal{EL}^{++}$ (with N_C replaced by $N_\mathrm{C} \cup N_\mathrm{N}$), as the occurrence of inconsistencies generally reduces the number of paths in $\mathcal{H}_{\mathbf{RAP}}$. This is summed up in the following corollary.

Corollary 3.20 (Correctness and complexity of RAPsolve-EL++)
Algorithm RAPsolve-EL++ is correct and terminates with runtime

$$O\left(2^{|\mathcal{A}|+|\mathcal{O}|} \cdot |N_\mathrm{C}|^2 \cdot |N_\mathrm{R}| + 2^{|\mathcal{A}|+|\mathcal{O}|} \cdot \left(T_{\prec_\mathcal{A}}(|\mathcal{A}|) + T_{\prec_\mathcal{O}}(|\mathcal{O}|)\right)\right).$$

During its execution, pruning based on $\subseteq$ can be applied without changing the solution set, whereas pruning based on $\models$ and $\models_\mathcal{T}$ yields potentially smaller but logically equivalent solution sets. If subset-based pruning is employed, the runtime is reduced to

$$O\left(2^{|\mathcal{A}|} \cdot |N_\mathrm{C}|^2 \cdot |N_\mathrm{R}| + 4^{|\mathcal{A}|} \cdot \left(T_{\prec_\mathcal{A}}(|\mathcal{A}|) + T_{\prec_\mathcal{O}}(|\mathcal{O}|)\right)\right)$$

in the worst case, respectively

$$O\left(1.5^{|\mathcal{A}|} \cdot |N_\mathrm{C}|^2 \cdot |N_\mathrm{R}| + 2.25^{|\mathcal{A}|} \cdot \left(T_{\prec_\mathcal{A}}(|\mathcal{A}|) + T_{\prec_\mathcal{O}}(|\mathcal{O}|)\right)\right)$$

in average.

Relaxed Abduction over Horn$-\mathcal{SHIQ}$ **Knowledge Bases.** Another prominent example of a description logic that allows for consequence-driven reasoning mechanisms is Horn$-\mathcal{SHIQ}$ (Kazakov, 2009). Its complete set of language features includes concept conjunction ($\mathsf{C} \sqcap \mathsf{D}$), disjunction ($\mathsf{C} \sqcup \mathsf{D}$), and negation ($\neg\mathsf{C}$) as well as qualified existential ($\exists\,\mathsf{r}\,.\,\mathsf{C}$), value ($\forall\,\mathsf{r}\,.\,\mathsf{C}$) and cardinality restrictions ($\geq m\,\mathsf{r}\,.\,\mathsf{C}$, $\leq n\,\mathsf{r}\,.\,\mathsf{C}$). Different from the $\mathcal{EL}$ family but similarly to the OWL 2 RL profile of the Web Ontology Language (W3C OWL Working Group, 2009b), Horn$-\mathcal{SHIQ}$ provides asymmetric expressiveness for sub- and superconcepts of a concept inclusion axiom.

Algorithm 3.16: RAPsolve-EL++

Input: RAP $= (\mathcal{T}, \mathcal{A}, \mathcal{O}, \preceq_{\mathcal{A}}, \preceq_{\mathcal{O}})$ $- an$ $\mathcal{EL}^{++}$**-RAP** $over$ $N_C^\top$ and N_R
Output: $Sol_{\mathbf{RAP}}$

```
   // initialization
1  No ← ∅;
2  foreach r ∈ N_R do
3      R(r) ← ∅;
4  foreach C ∈ N_C^⊤ ∪ N_N do
5      L ← ∅;
6      if C ⊑ C ∈ 𝒪 then  L ← L ∪ {C : {(∅, C ⊑ C)}};
7      else  L ← L ∪ {C : {(∅, ∅)}};
8      if C ⊑ ⊤ ∈ 𝒪 then  L ← L ∪ {⊤ : {(∅, C ⊑ ⊤)}};
9      else  L ← L ∪ {⊤ : {(∅, ∅)}};
10     S(C) ← S(C) ∪ L;
   // propagation
11 repeat
12     changed ← false;
13     foreach ax ∈ 𝒯 ∪ 𝒜 do
14         switch typeof(ax) do
15             case (NF1) changed ← changed | applyCR1(ax);
               // (...)  other cases c.f. RAPsolve-EL+
21     changed ← changed | applyCR7();
22     changed ← changed | applyCR8();
23 until changed = false;
   // solution gathering
24 Sol ← {(∅, ∅)};
25 foreach C1 ⊑ D ∈ 𝒪 do
26     if S(C1) ∋ D : L then  Sol ← join-and-prune(Sol,L);
27     if S(C1) ∋ ⊥ : L then
28         Sol ← join-and-prune(Sol,meet(L,{(∅,∅)},null,C1 ⊑ D));
29 foreach C1 ⊑ ∃r1.C2 ∈ 𝒪 do
30     if R(r1) ∋ (C1, C2) : L then  Sol ← join-and-prune(Sol,L);
31 return meet-closure(Sol);
```

Procedure 3.17: `applyCR1'` (with nogood handling)

Input: ax $= C1 \sqsubseteq D - a$ *(NF1)-axiom*
Output: **true** if the rule application was productive, **false** otherwise

1   changed $\leftarrow$ **false**;
2   **foreach** $C \in N_C^\top \cup N_N$ **do**
3   **if** $S(C) \ni C1 : L1$ **then**
        // (...) determine L, Lnew as in applyCR1
7   **if** Lnew $\neq$ L **then**
8   **if** $D = \bot$ & $C \in N_N$ **then** add-nogoods(Lnew);
9   **else** $S(C) \leftarrow S(C) \setminus \{D : L\} \cup \{D : \text{Lnew}\}$;
10   changed $\leftarrow$ **true**;

11   **return** changed;

Procedure 3.18: `applyCR2'` (with nogood handling)

Input: ax $= C1 \sqcap C2 \sqsubseteq D - a$ *(NF2)-axiom*
Output: **true** if the rule application was productive, **false** otherwise

1   **foreach** $C \in N_C^\top \cup N_N$ **do**
2   **if** $S(C) \ni C1 : L1 \wedge S(C) \ni C2 : L2$ **then**
        // (...) determine L, Lnew as in applyCR2
7   **if** Lnew $\neq$ L **then**
8   **if** $D = \bot$ & $C \in N_N$ **then** add-nogoods(Lnew);
9   **else** $S(C) \leftarrow S(C) \setminus \{D : L\} \cup \{D : \text{Lnew}\}$;
10   changed $\leftarrow$ **true**;

11   **return** changed;

Procedure 3.19: `applyCR4'`   (with nogood handling)

Input: $\text{ax} = \exists r1.C2 \sqsubseteq D - a$ *(NF4)-axiom*
Output: **true** if the rule application was productive, **false** otherwise

1 **foreach** $C1 \in N_C^\top \cup N_N$ **do**
2 | **if** $S(C1) \ni C2 : L1$ **then**
3 | | **foreach** $C \in N_C^\top \cup N_N$ **do**
4 | | | **if** $R(r1) \ni (C, C1) : L2$ **then**
   | | | | `// (...) determine L, Lnew as in applyCR4`
9 | | | | **if** $\text{Lnew} \neq L$ **then**
10 | | | | | **if** $D = \bot$ & $C \in N_N$ **then** `add-nogoods(Lnew)`;
11 | | | | | **else** $S(C) \leftarrow S(C) \setminus \{D : L\} \cup \{D : \text{Lnew}\}$;
12 | | | | | changed $\leftarrow$ **true**;

13 **return** changed;

Procedure 3.20: `applyCR7`   (with nogood handling)

Output: **true** if the rule application was productive, **false** otherwise

1 **foreach** $C1 \in N_C^\top \cup N_N$ **do**
2 | **if** $S(C1) \ni \bot : L1$ **then**
3 | | **foreach** $C \in N_C^\top \cup N_N$ **do**
4 | | | **foreach** $r1 \in N_R$ **do**
5 | | | | **if** $R(r1) \ni (C, C1) : L2$ **then**
6 | | | | | **if** $S(C) \ni \bot : \text{Lold}$ **then** $L \leftarrow \text{Lold}$;
7 | | | | | **else** $L \leftarrow \{(\emptyset, \emptyset)\}$;
8 | | | | | $\text{Lnew} \leftarrow \text{join}(L, \text{meet}(L1, L2, \text{null}, C \sqsubseteq \bot))$;
9 | | | | | **if** $\text{Lnew} \neq L$ **then**
10 | | | | | | **if** $C \in N_N$ **then** `add-nogoods(Lnew)`;
11 | | | | | | **else** $S(C) \leftarrow S(C) \setminus \{\bot : L\} \cup \{\bot : \text{Lnew}\}$;
12 | | | | | | changed $\leftarrow$ **true**;

13 **return** changed;

Procedure 3.21: `applyCR8`   (with nogood handling)

Output: **true** if the rule application was productive, **false** otherwise

1 **foreach** $\{a\} \in N_N$ **do**
2   **foreach** $C1 \in N_C^\top \cup N_N$ **do**
3   **if** $S(C1) \ni \{a\} : L1$ **then**
4   **foreach** $C2 \in N_C^\top \cup N_N$ **do**
5   **if** $S(C2) \ni \{a\} : L2$ **then**
6   **foreach** $r1 \in N_R$ **do**
7   **if** $R(r1) \ni (C1, C2) : L3$ **then**
8   **foreach** $D \in N_C^{\top,\perp} \cup N_N$ **do**
9   **if** $S(C2) \ni D : L4$ **then**
10   **if** $S(C1) \ni D : Lold$ **then** $L \leftarrow Lold$;
11   **else** $L \leftarrow \{(\emptyset, \emptyset)\}$;
12                            $Lnew \leftarrow$ `join(`L,
                                 `meet(`$L1$, $L2$,$L3$,$L4$,**null**,$C1 \sqsubseteq D$)`)`;
13   **if** $Lnew \neq L$ **then**
14   **if** $D = \perp$ & $C1 \in N_N$ **then**
15   `add-nogoods(`$Lnew$`)`;
16   **else**
17                                    $S(C1) \leftarrow S(C1) \setminus \{D : L\}$
                                   $\cup\{D : Lnew\}$;
18                            $changed \leftarrow$ **true**;

19 **return** $changed$;

Procedure 3.22: `add-nogoods`

Input: L *– a label set containing nogoods*

1 **foreach** $(A, O) \in L$ **do**
2   **foreach** $A1 \in No$ **do**
3   **if** $A \subset A1$ **then** $No \leftarrow No \setminus \{A1\}$;
4   **else if** $A \supset A1$ **then** **break**;
5        $No \leftarrow No \cup \{A\}$

6 **return**;

Function 3.23: `join-and-prune''`   (safe, with nogood handling)

Input: L1, L2 – *two label sets, the first is assumed dominance-free*
Output: L1 ⊕ L2 – *with all inclusion-dominated elements removed*

```
   // (...) determine L as in join-and-prune'
10 foreach (A, O) ∈ L do
11    foreach A1 ∈ No do
12       if A1 ⊆ A then
13          L ← L \ {(A, O)};
14          break;

15 return L;
```

Function 3.24: `remove-dominated'`   (with nogood handling)

Input: L – *a label set*
Output: The label set L with all ≺-dominated elements removed

```
  // (...) determine L as in remove-dominated
5 foreach (A, O) ∈ L do
6    foreach A1 ∈ No do
7       if A1 ⊆ A then
8          L ← L \ {(A, O)};
9          break;

10 return L;
```

Any Horn−$\mathcal{SHIQ}$ knowledge base can be classified based on a set of six completion rules, the consequences derived by the rules are of one of the forms $\bigsqcap A_i \sqsubseteq B$ and $\bigsqcap A_i \sqsubseteq \exists r. (\bigsqcap B_j)$.

Our completion-based solution for solving $\mathcal{EL}^{++}$ relaxed abduction problems can be extended to Horn−$\mathcal{SHIQ}$ by modifying the maps S and R: Instead of assigning concepts to concepts and pairs of concepts to roles, the modified mappings are defined by $S : \mathcal{P}(N_C^\top) \mapsto N_C^{\top,\perp}$ and $R : N_R \mapsto \mathcal{P}(N_C^\top) \times \mathcal{P}(N_C^{\top,\perp})$. The notion of a label is applicable without modification to these extended map entries; and inconsistencies are similarly handled by blocking the corresponding labels. The completion rules defined in Kazakov (2009) can then be employed similarly to propagate labels and thereby solve the **RAP**.

3.4.2 Incremental Relaxed Abduction

A second extension we consider in this section is the processing of changes to the underlying formalisation, i. e. the addition and retraction of axioms to the knowledge bases. From an application point of view, efficient handling of such modifications is essential in a number of domains including but not limited to diagnostics, where observational data is a prominent source for changes (see the following section on the dimensions of incrementality for a more detailed elaboration on this topic).

Obviously, a straightforward approach to addressing knowledge base updates is to restart the algorithm from scratch with the modified axiom sets $\mathcal{T}$, $\mathcal{A}$ and $\mathcal{O}$, therefore determining the results for the updated problem from scratch. This section investigates how this problem can be solved more efficiently by maximizing the reuse of available information. Incremental reasoning and particularly incremental classification of description logic knowledge bases have been studied before, most relevant for our approach being the paper of Grau et al. (2010): The authors propose both a module-based approach for general description logics that can handle addition and retraction of axioms, and a specialized algorithm for incremental addition of axioms to $\mathcal{EL}^+$ knowledge bases. While our approach is significantly more complex due to the fact that we use multiple knowledge bases with different roles, some of the ideas presented there can be carried over into our framework.

Dimensions of Incrementality

We start our investigation of incremental relaxed abduction by motivating the different dimensions of incrementality studied here. In concert with Grau et al. (2010) we investigate both the addition and the retraction of axioms. Different from their work, our approach is faced with three sets of axioms with different meaning to the problem and therefore requiring different algorithmic treatment.

Updating observations. The presumably most natural source for incrementality presented in Example 3.6 below is the observation set $\mathcal{O}$, as new observations are constantly made in most applications, and thus need to be integrated into the information process (we use the notation $\mathcal{O}+$ to refer both to the process of adding an observation axiom, and for the resulting updated observation set). Analogously observations may be invalidated, for example due to newer observations from the same source, or to some user-implemented deprecation process that automatically removes observations exceeding a given expiry age. Similarly, we abbreviate this retraction and the resulting axiom set by $\mathcal{O}-$.

Example 3.6 (Incrementality in diagnostics (updating $\mathcal{O}$))
In the context of the diagnostic example introduced in Example 3.1 on page 43, new sensor data provided by the system after the initial run might indicate that the gripper has stopped functioning as intended. From a theoretical perspective, incorporating this new information implies a belief revision problem in the sense of Halaschek-Wiener & Katz (2006): Since the intended statement $\neg$(Gripper $\sqsubseteq \exists$ shows . FullyFunctional) cannot be expressed in any description logic, the standard solution consists of removing all axioms from $\mathcal{O}$ that are incompatible with $\neg$(Gripper $\sqsubseteq \exists$ shows . FullyFunctional). However, for the sake of simplicity, we follow a more hands-on type of strategy by removing the observation Gripper $\sqsubseteq \exists$ shows . FullyFunctional *from* $\mathcal{O}$, *followed by the addition of* Gripper $\sqsubseteq \exists$ shows . IrregularMovements. *Note that since not all observations (that might potentially be incompatible with the intended axiom) need to be explained, this does not render the problem inconsistent. This changes both non-trivial solutions (A_1, O_1) and (A_2, O_2) identified previously into (A_1, O_1') and (A_2, O_2') as follows:*

$$A_1 = \{\text{System} \sqsubseteq \exists \text{ operatesIn . PowerSupplyFluctuations}\}$$
$$O_1' = \{\text{MCU} \sqsubseteq \exists \text{ shows . IntermittentOutages}\}$$
$$A_2 = \{\text{System} \sqsubseteq \exists \text{ operatesIn . ControlSWMalfunction}\}$$
$$O_2' = \{\text{MCU} \sqsubseteq \exists \text{ shows . IntermittentOutages},$$
$$\text{Gripper} \sqsubseteq \exists \text{ shows . IrregularMovements}\}$$

Note that when using inclusion-based dominance, both candidates still represent valid solutions. However, if cardinality-based dominance had been employed instead, the previously dominated (A_2, O_2) would be part of the new solution (A_2, O_2'), whereas the previously dominant (A_1, O_1) would be dominated after the update. Moreover, the subsequent addition of another observation PROFINET $\sqsubseteq \exists$ shows . SendingReceivingOK *would lead to an extended observation set $Sol_{\mathbf{RAP}} = \{(\emptyset, \emptyset), (A_1, O_1''), (A_2, O_2'), (A_1 \cup A_2, O_1'' \cup O_2')\}$ where the modified observation set O_1'' is given by:*

$$O_1'' = \{\text{MCU} \sqsubseteq \exists \text{ shows . IntermittentOutages},$$
$$\text{PROFINET} \sqsubseteq \exists \text{ shows . SendingReceivingOK}\}$$

Updating the domain ontology. The addition $(\mathcal{T}+)$ and retraction $(\mathcal{T}-)$ of axioms presented in Example 3.7 below to/from $\mathcal{T}$ is just the type of incremcntality considered by Grau et al. (2010). Reasons for such a modification might be slight updates of single axioms in a long-running information interpretation system, or the addition (or retraction) of larger portions of the knowledge base initiated by the user.

Example 3.7 (Incremenatlity in diagnostics (updating $\mathcal{T}$))
In interactive diagnostics, a user might decide to focus on a subsystem presumably causing the symptoms initially, using a less detailed representation of the machine ($\mathcal{T}' \subseteq \mathcal{T}$). If she finds that no satisfying diagnosis can be found this way, other subsystems can be added in a step-by-step process until certain quality criteria for a good solution are met. In our scenario, the ontology $\mathcal{M}_{sys}$ of the production system might be extended with a description of the WLAN component (WLAN $\sqsubseteq \exists$ belongsTo . Communications), including an extension of $\mathcal{M}_{caus}$ with the additional information that WLAN breakdowns are associated with fluctuations in the power supply, expressed by:

$$\text{WLAN} \sqcap \exists \text{ partOf} . (\exists \text{ operatesIn} . \text{PowerSupplyFluctuations})$$
$$\sqsubseteq \exists \text{ shows} . \text{Breakdown}$$

Updating abducibles. In principle, abducibles play a similar role to axioms from $\mathcal{T}$ from a technical point of view, they do however represent distinguished axioms whose use is to represent the solutions. Therefore, the motivation for removing ($\mathcal{A}-$) or adding ($\mathcal{A}+$) abducibles is similar to the argumentation given before. We prcscnt an example next.

Example 3.8 (Incremenatlity in diagnostics (updating $\mathcal{A}$))
In an interactive diagnostic process, the operator might decide to rule out certain diagnoses a priori to save runtime. She can do this by removing corresponding axioms from the set of abducibles. If her intuition proves wrong, the axioms can be re-inserted and the solution set updated accordingly. Assuming that in our example scenario the operator is not satisfied with the solution set derived so far, she might add a completely new diagnosis candidate (hypothesis) to $\mathcal{M}_{diag}$ (e. g. System $\sqsubseteq \exists$ operatesIn . BearingOutOfRound), followed by the addition of a definition of its effects devised by the domain experts to $\mathcal{M}_{diag}$. On the next run, given appropriate observations in $\mathcal{O}$ the diagnostic component will then display indications for a dysfunctional conveyor belt bearing as well.

Axiom migration. An experienced user using an information interpretation system may decide that an axiom from $\mathcal{T}$ should better be considered

abducible, or the other way round. Such a migration can be accomplished
straightforwardly by two consecutive operations of removing and (re-)adding
the axiom. We analyse for both algorithms proposed in Section 3.3 whether
such migrations (denoted $\mathcal{T} \rhd \mathcal{A}$ and $\mathcal{A} \rhd \mathcal{T}$) can be handled more efficiently
than by chaining $\mathcal{T} - \mathcal{A}+$ and $\mathcal{A} - \mathcal{T}+$ respectively.

Example 3.9 (Incremenatlity in diagnostics (axiom migration))

*In the diagnostics scenario, one sometimes has to consider systems which
are known to work in a suboptimal state (i. e. in some given failure mode),
and try to detect further faults in the execution of this partially working
system. In such a situation, the operator may save effort by moving the
known diagnoses from $\mathcal{A}$ to $\mathcal{T}$, so the system saves resources for tracking
the effects of other, unknown diagnoses. When the system is fully functional
again (e. g. after a maintenance operation), the respective axioms must then
be migrated back from $\mathcal{T}$ to $\mathcal{A}$.*

Incrementality in RAPsolve-Generic

We start our analysis of incrementality with the black-box algorithm for
solving relaxed abduction. The generic variant uses a boilerplate description
logic reasoner as a black-box system for classifying the knowledge bases
$\mathcal{T} \cup \mathcal{A}$. It does not have any access to the internal structures of the reasoner,
affecting the extent to which incrementality can be supported. However, the
unpruned set *Sol* contains a significant amount of reusable information. To
make this information accessible to the algorithm, we modify the loop in
line 9 of Algorithm 3.1 to not remove dominated elements from *Sol* but only
mark them accordingly.[15] This maximises information reuse at the cost of
losing the domination-induced reduction of the solution set size, typically a
factor of $O\left(\sqrt{|\mathcal{A}|}\right)$ (c.f. Propositions 3.4, 3.6 and 3.8). In the following, we
distinguish updated axioms sets (or relaxed abduction problems) from the
original ones by appending a prime to the name.

Addition $\mathcal{T}+$. RAPsolve-Generic never processes $\mathcal{T}$ directly, but passes it
to the deductive reasoning component as a whole. The addition of $\mathcal{T}$-axioms
must therefore be handled by re-classifying the knowledge bases $\mathcal{T}' \cup \mathcal{A}$.
Even if the deductive reasoner supports incremental classification, space
requirements are typically prohibitive for storing all combinations $\mathcal{T} \cup \mathcal{A}$
and incrementally adding the new $\mathcal{T}$-axiom(s) to each of them. It is however

---

[15]Note that two tuples (A, O_1) and (A, O_2) in $Sol_{\mathbf{RAP}}$ with the same assumption set
are still merged into one solution candidate $(A, O_1 \cup O_2)$.

possible to classify $\mathcal{T}'$ alone, and incrementally add $\mathcal{A}$-axioms to realise the iteration over the subsets $A \subseteq \mathcal{A}$ (either using axiom retraction, or by using several copies of the pre-classified knowledge base $\mathcal{T}'$). Independent from the underlying deductive reasoning component, information on inconsistency can be reused: If $A \sqsubseteq \mathcal{A}$ was inconsistent w. r. t. $\mathcal{T}$ (i. e. $A \cup \mathcal{T} \models \bot$), it must also be inconsistent w. r. t. the extended axiom set $\mathcal{T}'$ due to the monotonicity of entailment. Therefore, consistency only has to be re-evaluated for sets $A \sqsubseteq \mathcal{A}$ consistent with $\mathcal{T}$, i. e. those A that have a (dominated or undominated) entry $(A, O) \in Sol_{\mathbf{RAP}}$.[16] Similarly, monotonicity of entailment guarantees that $\mathcal{T} \cup A \models \{o\}$ implies $\mathcal{T}' \cup A \models \{o\}$. Therefore, the loop in line 5 of Algorithm 3.1 may update existing tuples (A, O) by iterating over $\mathcal{O} \setminus O$ only and adding the newly entailed observations. Note that $Sol_{\mathbf{RAP}}$ never grows in this process, tuples can only be removed (due to inconsistency) or have their O component expanded. It is however necessary to re-determine the dominance flags, since dominating entries may be removed, making the dominated elements optimal again.

Retraction $\mathcal{T}-$.   The situation of removing $\mathcal{T}$-axioms is symmetric to the update $\mathcal{T}+$ considered before: Again, a classifier supporting the incremental retraction of axioms is of limited use since the large number of $\mathcal{A}$-subsets prohibits an explicit storage; it can however be used to pre-classify $\mathcal{T}'$. Nevertheless, monotonocity of entailment can be employed analogously to the preceding case for reducing the number of entailment tests in line 6 of Algorithm 3.1 by only testing the elements in O for all $(A, O) \in Sol_{\mathbf{RAP}}$, and removing from them observations that are not longer entailed. Analogously, no new inconsistencies can occur and it therefore suffices to check $\mathcal{T} \cup A \models \bot$ for all $A \subseteq \mathcal{A}$ that were inconsistent before (i. e. that are not in $Sol_{\mathbf{RAP}}$ in marked or unmarked form). If a new inconsistent $A \subseteq \mathcal{A}$ is found, the explained observations are determined as usual. In this case, Sol may never lose any tuples (as no new inconsistencies can arise), but the O components of existing ones may shrink up to being empty. Dominance between the elements in $Sol_{\mathbf{RAP}}$ thus has to be re-established just like before.

Addition $\mathcal{A}+$.   Adding a new abducible to $\mathcal{A}$ doubles the number of subsets A that can potentially constitute a solution (for each subset A, the new axiom a can either be a member or not). Note however that the cases where $a \notin A$ are exactly the original sets $A \subseteq \mathcal{A}$ analysed already and

---

[16]This is made possible by *not* removing dominated entries but only marking them accordingly, as proposed before.

represented in $Sol_{\mathbf{RAP}}$, they can therefore be kept without modification. For determining the new tuples (A', O'), the information stored in $Sol_{\mathbf{RAP}}$ can be reused to a significant extent by analysing the effects of adding a to $(A, O) \in Sol_{\mathbf{RAP}}$: Firstly, if A was inconsistent w.r.t. $\mathcal{T}$ before (i.e. there is *no* $O \subseteq \mathcal{O}$ such that $(A, O) \in Sol_{\mathbf{RAP}}$), this inconsistency naturally carries over to $A \cup \{a\}$. For all $(A, O) \in Sol_{\mathbf{RAP}}$, the addition of a to A may either lead to an inconsistency, help explain new observations, or none of the two (in this case, the resulting tuple $(A \cup \{a\}, O)$ is obviously dominated by (A, O)). We therefore first test whether $A \cup \{a\}$ is consistent with $\mathcal{T}$, and only then determine the set of entailed observations (which must obviously be a superset of O, allowing us to only test the elements in $\mathcal{O} \setminus O$ for entailment), eventually adding the resulting solution candidate $(A \cup \{a\}, O')$ to $Sol_{\mathbf{RAP}}$. Finally, dominance between the solution candidates is determined.

Retraction $\mathcal{A}-$. For the removal of abducible a from $\mathcal{A}$, observe that all sets $A \subseteq \mathcal{A} \setminus \{a\}$ are necessarily also subsets of the original abducible set $\mathcal{A}$, and moreover any $A \subseteq \mathcal{A}$ that does not contain a remains unchanged. Therefore, no new $\mathcal{T}$-consistent subsets of $\mathcal{A}$ can emerge since $A \setminus \{a\}$ must already have a (possibly dominated) entry in $Sol_{\mathbf{RAP}}$ if it is consistent. It is therefore sufficient to remove from $Sol_{\mathbf{RAP}}$ all tuples (A, O) such that $A \ni a$, and re-calculate dominance among the remaining elements (since some of them may have become undominated in this process).

Migration $\mathcal{T} \rhd \mathcal{A}$. Note that removing an axiom from $\mathcal{T}$ and adding it to A changes neither consistency nor the set of entailed observations of a given tuple (A, O): $\mathcal{T} \cup A \models O$ if and only if $(\mathcal{T} \setminus \{a\}) \cup (A \cup \{a\}) \models O$ (for $a \in \mathcal{T}$). We can therefore realise the migration $\mathcal{T} \rhd \mathcal{A}$ in three consecutive steps: For each original $(A, O) \in Sol_{\mathbf{RAP}}$, we first add $(A \cup \{a\}, O)$ to $Sol_{\mathbf{RAP}}$ and then update the original entry (A, O) by removing all $o \in O$ that are no longer entailed after removing a from $\mathcal{T}$ (using the loop starting in line 9 of Algorithm 3.1, restricted to $\mathcal{O} \cup O$). Third, we determine for all $A \subseteq \mathcal{A}$ having no Sol entry (exactly the sets $A \subseteq \mathcal{A}$ where $A \cup \mathcal{T} \models \bot$) whether they are consistent w.r.t. the updated knowledge base $\mathcal{T}'$. If so, the entailment-checking loop of Algorithm 3.1 is used again to determine the set of observations O entailed by $\mathcal{T} \cup A$, and (A, O) is added to $Sol_{\mathbf{RAP}}$. Finally, dominance between the entries must be redetermined.

Migration $\mathcal{A} \rhd \mathcal{T}$. Note that conversely to the situation $\mathcal{T} \rhd \mathcal{A}$ analysed before, by turning an assumption into a axiom from $\mathcal{T}$ no new consistent solution candidates may occur; existing tuples (A, O) may however become inconsistent if A does not contain the migrated axiom a. It therefore suffices to iterate over the elements of $Sol_{\mathbf{RAP}}$, testing for each (A, O) whether $a \in A$. If the solution candidate uses axiom a it cannot become inconsistent by the migration, we can therefore simply remove the assumption a. If $a \notin A$, then (A, O) is removed from $Sol_{\mathbf{RAP}}$ if $\mathcal{T} \cup \{a\} \cup A \models \perp$, and otherwise we use the loop in Algorithm 3.1 to determine the observations entailed by $\mathcal{T} \cup \{a\} \cup A$, and add the resulting tuple to $Sol_{\mathbf{RAP}}$. After re-establishing dominance, the migration is completed.

Addition $\mathcal{O}+$. A new observation o can be integrated straightforwardly by iterating over all $(A, O) \in Sol$ and determining whether $\mathcal{T} \cup A \models \{o\}$, making use of a pre-classified knowledge base $\mathcal{T}$ if supported by the reasoner. If a previously dominated tuple (A, O) gets updated to $(A, O \cup \{o\})$ in this process, its dominance flag must be recomputed since the extension of O may render the new tuple $\preceq$-minimal.

Retraction $\mathcal{O}-$. An observation o can be retracted by truncating all tuples $(A, O) \in Sol_{\mathbf{RAP}}$ to $(A, O \cap \mathcal{O}')$. As this may reduce the explanatory power of solution candidates, the dominance flags of all updated tuples that were dominant (i. e. $\preceq$-minimal) before must be recomputed.

Incrementality in `RAPsolveEL+` and `RAPsolveEL++`

More efficient approaches to handling incrementality are to be expected for the completion-based variants (Algorithms 3.3 and 3.16) since all reasoner-internal data structures are accessible. Therefore, there is a higher potential for optimising such operations. Like before, we denote updated sets by the primed name of the original set. Moreover, we assume labels of mapping entries in S and R are either unpruned, or pruned based on set inclusion (c.f. Proposition 3.18). As before we do not realise pruning by actually removing the label elements, we rather use a flag to denote whether an entry is dominant (and thus a solution), or pruned (i. e. dominated). This implies that for all sets $A \subseteq \mathcal{A}$, A is consistent with $\mathcal{T}$ if and only if there exists $O \subseteq \mathcal{O}$ such that $(A, O) \in Sol_{\mathbf{RAP}}$. This guarantees that all information on derivations is available from the labels.

Addition $\mathcal{T}+$. Following the $\mathcal{EL}^+$-related results in Grau et al. (2010), a new normal form axiom can be incorporated into $\mathcal{T}$ by feeding it into the propagation loop (i. e. executing the appropriate `applyCRi` procedure),

and continuing propagation until saturation is reached again. As before
this update may lead to existing solutions being improved (by making an
assumption superfluous), while in case of $\mathcal{EL}^{++}$ other $Sol_{\mathbf{RAP}}$ members
may be rendered inconsistent w. r. t. $\mathcal{T}'$ and therefore need to be removed.
For speeding up the recomputation of Sol from the structures S and R,
we make advantage of this fact as follows: Once the label set of a vertex
corresponding to an observation $o \in \mathcal{O}$ is updated[17], the updated label set
is immediately joined with $Sol_{\mathbf{RAP}}$. Similarly, any new entry to the label of
the contradiction node ($\perp$) is immediately incorporated into the nogood set
No, and $Sol_{\mathbf{RAP}}$ is pruned accordingly. When propagation has terminated,
the meet-closure of Sol is redetermined (which can obviously be omitted
if no relevant label was updated).

Retraction $\mathcal{T}-$. Since Algorithm 3.3 does not store any information on the
$E^{\mathcal{T}}$-edges used, there is no way of retracting $\mathcal{T}$-axioms but restarting the
propagation from scratch for the updated problem $\mathbf{RAP}' = (\mathcal{T}', \mathcal{A}, \mathcal{O}, \preceq_{\mathcal{A}}$
$, \preceq_{\mathcal{O}})$, and thus for the reduced axiom set $\mathcal{T}' \cup \mathcal{A}$. Note that Algorithm 3.3
could easily be extended to store information on $E^{\mathcal{T}}$-edges as well by ex-
tending the labels to triples accordingly. However, the resulting increase in
label size and thus in runtime (proportional to $2^{|\mathcal{T}|}$) is prohibitive already
for mid-sized domain ontologies.

Addition $\mathcal{A}+$. The only difference in Algorithm 3.3 between processing
axioms from $\mathcal{T}$ and $\mathcal{A}$ lies in the label processing. Therefore, the addition of
abducibles can be handled similarly to the update $\mathcal{T}+$ analysed before. Yet,
unlike the addition of $\mathcal{T}$-axioms, the addition of abducibles will never render
existing solutions inconsistent and existing solutions cannot be improved. It
may however give rise to new label entries, which can be incorporated into
$Sol_{\mathbf{RAP}}$ in a similar fashion as described for $\mathcal{T}+$[18].

Retraction $\mathcal{A}-$. Keeping in mind the intuition that the labels of entries
in S and R encode paths to the corresponding vertex in the hypergraph,
removal of a $E^{\mathcal{A}}$-edge can be realized by removing all paths (respectively
their representing label entries) that depend on this edge. This can easily
be done by iterating over the elements of S and R, and removing all label

---

[17]In typical applications, there will be few such successful updates for the addition of a
single axiom.

[18]In general, propagations are even more unlikely to be effective for this update, since
the addition of an assumption can only improve a solution of new observations are
entailed by it

entries (A, O) where $A \not\subseteq \mathcal{A}'$, that is, whose first component A contains the assumption to be removed. Similarly, the solution set $Sol_{\mathbf{RAP}}$ can be reduced this way my removing all entries (A, O) relying on the assumption to be removed due to the fact that consistency remains unchanged. Dominance then has to be redetermined for all $Sol_{\mathbf{RAP}}$ entries marked as dominated, since the dominating entry may have been removed.

Migration $\mathcal{T} \triangleright \mathcal{A}$. The induced hypergraph $\mathcal{H}_{\mathbf{RAP}}$ remains structurally unchanged by this operation since migrating axioms between $\mathcal{T}$ and $\mathcal{A}$ does not change the paths that can be found. However, path weights may require an update due to the fact that using the migrated axiom a is no longer "for free". Similarly to the update $\mathcal{T}-$ discussed before, this could in principle be supported efficiently by extending the labels with information on the $E^{\mathcal{T}}$-edges used. For similar reasons, we opt against such a modification, and re-calculate the completion graph from scratch here.

Migration $\mathcal{A} \triangleright \mathcal{T}$. Different from the migration discussed before, the structure of the induced hypergraph may change due to this update since using a as a $\mathcal{T}$-axiom may induce new inconsistencies. However, this can only happen for label entries (A, O) where $a \notin A$ since otherwise $\mathcal{T} \cup A \models \bot$. Note that this time information on the use of axiom a (this time as justification of an $E^{\mathcal{T}}$-edge) is encoded in the labels and can be used to avoid a complete recomputation: Firstly, we propagate the "new" axiom a using the standard algorithm. If new nogoods arise, they are removed from all node labels. Only then, all label entries (A, O) are updated to $(A \cap \mathcal{A}', O) = (A \setminus \{a\}, O)$ (it obviously suffices to restrict to entries where $a \in A$). In this process, within each label entries with identical first component are merged into one, replacing for example the two entries (A, O_1) and (A, O_2) by a single new entry $(A, O_1 \cup O_2)$. Finally, dominance is redetermined and $Sol_{\mathbf{RAP}'}$ is recomputed from the labels.

Addition $\mathcal{O}+$. For the addition of a new axiom a to $\mathcal{O}$, keep in mind that observations correspond to distinguished vertices in the induced hypergraph $\mathcal{H}_{\mathbf{RAP}}$. Therefore, only this very node and nodes using it as a premise either directly or indirectly can be affected by the addition. We start by looking up the map entry corresponding to a in S or R, depending on its type. If there is no entry, the vertex is not reachable from $V_{\mathcal{T}}$ and we are done, the new observation is not derivable at all. Otherwise each entry (A, O) of the label set of a is updated to $(A, O \cup \{a\})$; dominance is not affected by this update

	$\mathcal{T}+$	$\mathcal{T}-$	$\mathcal{A}+$	$\mathcal{A}-$	$'I \triangleright \mathcal{A}$	$\mathcal{A} \triangleright \mathcal{T}$	$\mathcal{O}+$	$\mathcal{O}-$
generic	◯[19]	◯[19]	◯[19]	◯[19]	◯[19]	◯[19]	◯[19]	✓
$\mathcal{EL}^{++}$	✓		✓	✓	✓	✓	✓	✓

Table 3.3: Potential for improvements in efficiency by reusing information in RAPsolve-Generic and RAPsolve-EL++

as all entries are extended identically. To spread the effects of this update through the hypergraph, the updated label is propagated along all outgoing ($E^{\mathcal{T}}$- and $E^{\mathcal{A}}$-) edges of the node for a by applying all axioms from $\mathcal{T}$ and $\mathcal{A}$ (where only propagations having a as a premise need to be considered). After propagation has terminated, $Sol_{\mathbf{RAP'}}$ is redetermined as usual.

Retraction $\mathcal{O}-$. Analogously, removing an observation $a \in \mathcal{O}$ corresponds to the vertex a losing its status as a designated observation node. Since this worsens the affected label entries, we cannot simply use the propagation approach proposed for $\mathcal{O}+$ as the updated entries are dominated by the original ones and propagation would therefore stop without effect. Instead, the labels of all elements of S and R are pruned to $(A, O \cap \mathcal{O}')$ accordingly, and dominance is redetermined (note that entries that were undominated before and have not been updated are still undominated and need not be rechecked). The same pruning process can be used to update $Sol_{\mathbf{RAP}}$ to $Sol_{\mathbf{RAP'}}$.

Taken together, Section 3.4.2 shows that both the black-box solution and the glass-box approach can be extended to handle incremental changes to the underlying knowledge bases to a certain extent. The glass-box algorithm is more flexible in this respect as it allows to access information on the concrete derivation step when an assumption was used, or an observation was explained. Nevertheless, modifications incurring the retraction of $\mathcal{T}$-axioms are hard in both approaches. The $\mathcal{EL}^+$-based solution additionally benefits from the fact that $\mathcal{EL}^+$ knowledge bases can never be inconsistent, alleviating the need to check for inconsistencies arising from the addition of axioms to $\mathcal{T}$ and $\mathcal{A}$. The results of this section are summarised in Table 3.3.

---

[19] reuse of information on consistency and entailed observations possible

3.5 Comparison to Alternative Approaches

In this section, we survey relevant related work in the topics touched by this thesis, and compare these approaches to the solution developed in this chapter. For better accessibility, we structure our analysis along the subjects tackled in our work, starting with alternative approaches for automatic information interpretation and diagnostics. We then turn to other applications of consequence-driven reasoning schemes, and conclude by surveying approaches in the field of truth maintenance, which can be seen as a core task in relaxed abduction in general, and in the incremental settings in specific. The results of our analysis are summarised in Table 3.4 at the end of this section.

3.5.1 Information Interpretation and Diagnostics

It has been argued that logic-based abduction and specifically weighted abduction is an appropriate tool for information interpretation tasks including, but not limited to, text analysis (Hobbs et al., 1993). The full range of minimal solutions (either w. r. t. set inclusion or w. r. t. weight) is explored to determine the best solution for a text analysis problem, and given the fundamental constraints on complexity of abduction (c.f. Eiter & Gottlob, 1995), this approach based on rules can be implemented efficiently. As we have shown in Section 2.2.3, such uni-criterial abductive approaches can however only address observational incompleteness adequately, whereas they are prone to fail with respect to incomplete or incorrect domain formalisations. Its applicability to the use cases considered here is therefore restricted. The same argument carries over to common abduction approaches for description logics such as Bienvenu (2008), or filtering by abduction as proposed in Baral (2000) which bears similarities to conditioning in Bayesian networks.

Reggia et al. (1983) propose set covering as a method for diagnostic problems. This approach tries to maximally cover the set of observations by manifestations of as few diagnoses as possible; it does neither require that all manifestations have been observed, nor that all observations are covered by a manifestation. Thereby, set-cover abduction addresses both types of incompleteness considered here. However, the class of domain formalisations supported is restricted to simple, explicit mappings between diagnoses and manifestations. The authors propose a sound and complete expert system based on their approach. Allemang et al. (1987) investigate the complexity of this approach, showing its NP-completeness. As we have pointed out earlier in this line, set-cover abduction bears a close relationship to model-based diagnosis, its extension towards so-called fault models, and (certain aspects of) qualitative reasoning (see De Kleer & Kurien, 2003; Struss, 1997). With set-cover abduction, these approaches share the restriction that the underlying domain representation does not provide a rich domain formalisation but a rather simple direct mapping from faults to effects (which is obviously a core feature in qualitative reasoning, where complex mathematical formulae modelling physical effects are replaced by qualitative statements about the behaviour of a system).

The media interpretation framework proposed by Peraldi et al. (2007, 2009) can be seen as a combination of description logics and abductive logic programming (Denecker & Kakas, 2002) addressing incompleteness of the underlying knowledge base. Abduction is implemented as query answering over assertional data based on a combination of backward chaining DL-safe rules for abduction (where missing premises are assumed) with a $\mathcal{SHIQ}$ knowledge base that is used to derive new entailments resulting from these assumptions. The explanation space is explored in a best-first manner by keeping the set of alternative worlds (represented as Aboxes) in a so-called agenda from which the most promising interpretation Abox is selected for further expansion, where "most promising" is based on a uni-criterion measure that rewards the use of asserted information while penalizing assumptions. Using clever heuristics for interestingness, the approach can successfully determine a single interpretation in reasonable runtime. However, the method used to determine the solution should be considered an engine rather than an algorithm in the strict sense of the word, since properties such as termination and correctness have not been formally investigated.[20] Building on these results, a probabilistic extension based on Markov Logic has been proposed (Gries et al., 2010; Nafissi, 2013). While bi-criteriality of information interpretation is taken into account here, the two-step approach for solution generation implicitly requires a separability of consilience and simplicity for the results to be correct (an assumption which does however not hold). Similar to its ancestor this approach lacks rigorous proofs of properties such as termination and completeness, making it an engine for the generation of solutions. Moreover, the use of Markov Logic adds significant computational complexity.

Abductive logic programming (ALP) complements a (typically definite Horn) logic program with two sets of integrity constraints (which can be

---

[20]Termination and correctness of approaches that combine forward and backward chaining of rules have been investigated, among others, by Baget et al. (2011). The major result is that both can be guaranteed only in selected cases where the interaction between forward and backward rules is significantly restricted.

understood to specify e. g. mutual exclusivity of assumptions) and open predicates (defining the space of abducibles), respectively. Using a goal-driven procedure based on minimal-model semantics, ALP can be used to derive a hypothetical explanation for one observation (Kowalski, 2011). Obviously, a trade-off between hypothesis complexity and expressivity is of little use in this situation since any candidate either matches the goal (i. e. perfect expressivity) or not (no expressivity at all). Similar to the approaches by Gries et al.; Nafissi; Peraldi et al., abductive logic programming combines forward chaining and backward chaining rule applications to derive explanations (or, similarly, to generate plans for achieving a goal). Different from these works, the formal framework seems to be better developed; however the approach does neither generate alternative solutions nor take qualitative information (such as probabilities) into consideration.

The Theorist framework (Poole, 1988) resembles our approach in the respect that it defines and implements hypothetical reasoning over formally represented domain models based on a user-defined set of hypotheses. Similarly to the solution we propose, the general theory does not pose restrictions on the logical formalism used for information representation, in principle permitting full first-order logic. Different from our approach, however, an explanation in the sense of Poole (1988) only justifies a single observation, and the concrete hypotheses (assumptions) forming the explanation must be ground. None of these two restrictions is present in the approach presented here, and it moreover allows to provide partial yet simple explanations in situations where a larger set of observations is available.

Abox abduction has been addressed in Klarmann et al. (2011) as well, where ideas from tableaux-based reasoning over $\mathcal{ALC}$ knowledge bases are combined with a first-order resolution calculus to determine abductive solutions to a query using a goal-directed exploration strategy. Similar to the approach by Peraldi et al. presented before, both the information to be explained and the abductive solutions are represented by means of Abox assertions. Different from it, a strict formulation of logic-based abduction is used which requires that all observations are entailed by a solution and makes this approach vulnerable to imperfections in the formalisation of the domain. The proposed sound and complete algorithm determines consistent, relevant and minimal solutions to the abduction problem; however neither complexity issues nor the problem of unexplainable observations are addressed by the authors.

A number of so-called graphical models for knowledge representation and reasoning can also be employed for diagnostic (or abductive) reasoning; examples include most probable explanation inference in Bayesian networks and abductive reasoning for Markov logic networks (Kate & Mooney, 2009). Whereas Bayes nets and other approaches building on them typically lack full relational expressiveness (especially in domains with an unknown number of individuals), Markov logic networks (MLNs) are claimed to unify first-order and probabilistic reasoning. Abduction in MLNs is accomplished by defining soft rules that reverse the causality of the original rules, designed to create only a single, best explanation for the complete observation set. Determining this single solution nevertheless incurs immense computational cost, typically leading to the use of approximate algorithms. We therefore consider Markov logic of limited use here. This is further strengthened by the fact that probabilistic approaches have higher information requirements (including, but not limited to, conditional probabilities, potentials, or rule weights) as compared to their deterministic counterparts, necessitating the use of learning strategies.

A more detailed survey of most of these these and several other methods used for knowledge-based diagnosis can be found in the paper of Dressler & Puppe (1999), where the approaches are evaluated with respect to a collection of functional criteria covering most of the requirements posed here. However, neither of these approaches provides the unique combination of features that relaxed abduction offers, therefore making them inferior to the solution presented here for the task at hand.

3.5.2 Consequence-Driven Reasoning Methods

As explicated in Section 3.3.2, consequence-driven reasoning for description logics has first been explored in the context of deductive inference, namely classification of $\mathcal{EL}$ knowledge bases. Since its introduction, various optimisations to the algorithm have been proposed (see Mendez et al., 2011); most of these adaptations are applicable to our algorithm straightforwardly. Up until now, this form of reasoning has however only been applied to classification and axiom pinpointing, a reasoning task that allows to determine minimal axiom sets that entail a certain conclusion (Baader et al., 2007). To our best knowledge, the research we conducted is the first concerned with applying consequence-driven reasoning to information interpretation, and especially abduction, to address the problems posed by incomplete information.

3.5.3 Truth Maintenance and Reasoning

A truth maintenance system (TMS) is a means for storing and manipulating facts along with their interdependencies; it is however *not* a reasoner in its own right (Forbus & De Kleer, 1993). Basically, a TMS can be understood as a graph whose vertices are propositions while the edges correspond to dependencies between these propositions. This construction allows to maximise information reuse when tracking and updating dependencies, and gives rise to approaches such as dependency-directed backtracking in case of inconsistencies. The information for building this structure is produced by an inference component whereas the TMS itself is blind to the semantics of the nodes, a solitary TMS can therefore be understood as an alternative representation of the mapping e from set-cover abduction (c.f. Section 2.2.1).

In the context of relaxed abduction, the induced hypergraph $\mathcal{H}_{\mathbf{RAP}}$ is both structurally and semantically closely related to an assumption-based truth maintenance system (De Kleer, 1986). Assumption-based TMS (or ATMS, for short) attribute nodes with contexts in which they are true, thereby making it possible to efficiently trace the truth of propositions upon addition and retraction of information in several scenarios in parallel. This reveals a close similarity between ATMS and (incremental) abduction, where the scenarios correspond to different sets of assumptions as explicated in Paul (1993). In this light, the algorithm RAPsolve-EL+ can be understood as a system integrating an ATMS with an inference engine much tighter than clause management systems (CMS) as proposed in Reiter & De Kleer (1988) do. Additionally, we are not aware of any extensions to the framework of truth maintenance systems that addresses the problem of inexplicable observations in general, let alone in a bi-criterion approach as developed in this thesis.

Table 3.4: Comparison of related approaches

Approach	R1	R2	R3	R4	R5	R6	R7
Baral (2000); Bienvenu (2008); Hobbs et al. (1993)	✓	✓		✓	○	✓	✓
Allemang et al. (1987); Reggia et al. (1983)		✓	✓	✓	○	✓	✓
Poraldi et al. (2007, 2009)	✓	✓	✓	✓		○	○
Gries et al. (2010); Nafissi (2013)	✓	✓	✓	✓	○	○	
Kowalski (2011)	✓	○		○		✓	○
Poole (1988)	✓	✓		○	○	✓	○
Klarmann et al. (2011)	✓	✓		✓		✓	○
Baader et al. (2007)	✓			✓		✓	✓
Forbus & De Kleer (1993); Paul (1993)	○	✓		✓		✓	✓

4 Case Studies and Evaluation

This chapter reports on the practical results of our research. We first present an overview of the RAbIT system, which we developed as library for solving relaxed abduction problems. RAbIT provides both a glass-box algorithm for $\mathcal{EL}^+$ knowledge bases, and a black-box variant with a broader range of supported representation languages. Next, we present a comparative performance evaluation between the two implementations based on an $\mathcal{EL}^+$-**RAP** adapted from a practical use case. We conclude the evaluation chapter of this thesis by presenting two concrete scenarios we applied relaxed abduction in, and reporting on the feedback from a small group of experts.

4.1 Implementation of the RAbIT System

We have implemented RAPsolve-Generic (Algorithm 3.1) and the completion-based variant RAPsolve-EL+ (Algorithm 3.3) in a library called RAbIT (for *Robust Abductive Inference Tool*) based on the Java 6 programming language and the OWL API (Horridge & Bechhofer, 2009). The OWL API is employed for loading the knowledge bases $\mathcal{T}$, $\mathcal{A}$, and $\mathcal{O}$ from OWL ontology files stored on the harddisk, and to provide an abstraction layer from the concrete description logic reasoner used as a black-box classification engine in the generic algorithm. For the generic implementation, any description logic reasoner providing an OWL API interface and supporting its isEntailed() method can be used.

Both our implementations closely realises the pseudocode introduced before. One notable implementation detail, however, is the reuse of labels to reduce memory consumption. For RAPsolve-EL+, our first, naïve implementation frequently ran into OutOfMemoryErrors for typical problem sizes, failing to realize the theoretical scalability results in practice. In a detailed analysis we found that the labels consumed most of the available memory, with garbage collection frequently kicking in to dispose of instances no longer active due to updates (addition of alternative derivations) or pruning. Especially in situations where there exists a large number of derivations (and, thus, labels), the garbage collector finally failed to free the required memory, leading to the observed OutOfMemoryError. We were able to significantly reduce memory consumption by combining the object pool pattern[1] with compact data structures: Instead of using a standard Java collection for representing axiom sets, bitvectors represent subsets of $\mathcal{A}$ and $\mathcal{O}$, respectively. Additionally, both the total number of UnaryLabel instances as well as the rate of object creation were reduced significantly by the introduction of a factory for UnaryLabel objects. Duplicate instances are thereby replaced by references to the same object, which is particularly effective in our settings as labels such as $(\emptyset, \emptyset)$ may occur hundreds or thousands of times. The factory class keeps track of the number of active references to a label. Modification of a UnaryLabel (like in the meet() method) is only permitted if there is at most one reference to the label object – otherwise, a the factory class returns a reference to another instance representing the modified label (which is created automatically if needed), and reduces the use count of the original label by one. Labels with a reference counter of 0 are made available for garbage collection. For the blackbox variant RAPsolve-Generic, we have adopted the compact representation of axioms sets to ensure comparability of the results although label overhead proved to be less of a problem here.

To validate the implementations, we have tested RAPsolve-Generic with Pellet (Sirin et al., 2007) version 2.3.0, and JCEL (Mendez, 2012) version 0.16.1, respectively, against our implementation of RAPsolve-EL+. We conducted back-to-back tests on three different relaxed abduction problems formulated in $\mathcal{EL}$ and $\mathcal{EL}^+$ respectively. In all test cases the instantiations of RAPsolve-Generic provided identical solution sets. Moreover, the solutions were consistent with the ones provided by our implementation of RAPsolve-EL+, the only difference being whether trivial tautologies such as $C \sqsubseteq C$ are listed in the explanation sets or not. We therefore conjecture that both algorithms have been implemented correctly.

4.2 Performance and Scalability

In this section, we analyse how both the black-box and the glass-box algorithm implemented in RAbIT scale for **sRAP**[2] with increasing problem size. To ensure comparability of the results we base our analysis on $\mathcal{EL}^+$ knowledge bases, using JCEL as a reasoning component for the black-box variant. JCEL is known to be among the most efficient reasoners for the

---

[1] http://en.wikipedia.org/wiki/Object_pool_pattern

[2] **sRAP** has been chosen as a very general, yet strict instantiation of the general framework.

EL family of description logics, and it furthermore follows a completion-based approach just like our glass-box algorithm, allowing for a unbiased comparison between the black-box and the glass-box implementation.

Recapitulating the complexity results devised in Sections 3.3.1 to 3.3.3, the number of abducibles is the primary factor determining the size of the solution set and consequently the runtime of both algorithms, whereas the influence of the number of observations is significantly smaller if pruning is employed). The size of the underlying knowledge base $\mathcal{T}$, then again, significantly affects the effort required for classification respectively the number of rule applications, contributing a polynomial factor to the overall complexity for the description logic $\mathcal{EL}^+$. We validate these theoretical results in this section by evaluating the runtime of the generic and the completion-based variant, both using subset inclusion as preference order and as pruning criterion. The scenario we use for this evaluation is based on the use case discussed in Examples 2.1 and 3.1. The corresponding $\mathcal{EL}^+$-relaxed abduction problem **RAP** is given by

- a knowledge base $\mathcal{T}$ encoding (a simplified variant of) the structural and diagnostic knowledge on a steam and gas turbine by means of 49 concept inclusion axioms and 2 role inclusion axioms that range over 27 concept names and 4 role names,

- a knowledge base $\mathcal{A}$ representing 4 possible diagnoses using one concept inclusion axiom for each, and

- knowledge base $\mathcal{O}$ representing potential observations by means of 90 concept inclusion axioms (one axiom per observation), of which 12 are actually entailed by $\mathcal{T} \cup \mathcal{A}$.

We have derived scaled versions from this base scenario by "cloning" the described system along with its observations and diagnoses. Using a scaling factor of 1 / 10 / 100 / 1000 for $\mathcal{T}$ (denoted xT from now on) and $\mathcal{O}$ (denoted xO), and 1 / 4 / 7 / 10 for $\mathcal{A}$ (denoted xA), this results in a total of 51 / 483 / 4803 / 48003 $\mathcal{T}$-axioms[3], 90 / 900 / 9000 / 90000 $\mathcal{O}$-axioms, and 4 / 16 / 28 / 40 $\mathcal{A}$-axioms. For each combination of scaling factors we conducted 20 runs of RAbIT, 10 using the implementation of RAPsolve-Generic, and 10 using RAPsolve-EL+. Runs not completed after 24 hours of runtime were considered failed and aborted. We take the average over the 10 runs each as basis to level out side effects.

Figure 4.1 depicts the relative runtime of the two algorithms for different scaling factors, with xA being depicted on the horizontal axis and either

---

[3] Among the 51 axioms in $\mathcal{T}$, 3 describe the general problem structure and are therefore not multiplied, whereas the remaining 48 express system-specific knowledge.

xT or xO represented as line type. The relative runtime is averaged over the values of the remaining scaling factor. That is, the line for $xT = 10$ plots the value of xA against the relative runtime for this scaling factor xA and xT fixed to 10, averaged over $xO \in \{1, 10, 100, 1000\}$. The plots in Figure 4.1(a) are consistent with the theoretical results: For increasing values of xA, RAPsolve-EL+ outperforms RAPsolve-Generic by several orders of magnitude due to the lower (expected) basis of 1.5 instead of 2. This clearly demonstrates the added value of providing a specialized implementation for lightweight description logics. The anomaly at $xA = 1$, where the blackbox approach slightly outperforms the glassbox one for $xT = 1$ can be explained by the fact that JCEL is highly optimized and therefore has less overhead than RAbIT. Figure 4.1(b) depicts the same ratio split by the scaling factor of $\mathcal{O}$ instead of $\mathcal{T}$. Again, the observed dominance of the specialized algorithm for growing number of hypotheses matches he theoretical results.

Next, we investigate whether the algorithm RAPsolve-EL+ implemented in RAbIT is fast enough to make relaxed abduction useful in practical contexts. In the next series of graphs (Figures 4.2 and 4.3), we depict concrete runtime (measured in seconds) for a number of experiments in order to answer this question. Note that the runtime is displayed in logarithmic scale. Therefore, the linear (or polynomial) dependency on xA in all four plots clearly shows the expected exponential influence of $|\mathcal{A}|$. In contrast, the dependency of the overall runtime on the size of $\mathcal{O}$ is minor except for the case of small $\mathcal{T}$, as can clearly be seen by comparing Figure 4.2(a) to Figure 4.2(b).Finally, the plots in Figure 4.3 show an almost quadratic dependency between the size of $\mathcal{T}$ and the total runtime, matching well-known results on the complexity of reasoning in $\mathcal{EL}^+$. What do those numbers mean for the applicability of relaxed abduction in real-world use cases? We will address this question with two concrete examples in the upcoming section. Yet, the limitations of the proposed approach become clearly visible in the charts: Whereas the evaluation run took (on average) less than a second for the base **RAP** having $|\mathcal{T}| = 51, |\mathcal{A}| = 4, |\mathcal{O}| = 12$, already the scaled version with $|\mathcal{T}| = 4803, |\mathcal{A}| = 16, |\mathcal{O}| = 9000$ already took 1.2 hours to complete. Test cases with $xA = 7$ ($|\mathcal{A}| = 16$) could only be solved within our 24 hour limit if $xT <= 10$, for instance a run with $|\mathcal{T}| = 483, |\mathcal{A}| = 28, |\mathcal{O}| = 9000$ took on average 1.2 hours as well. While memory usage has not been in the scope of our evaluation, it is worthwhile to mention that our implementation of RAPsolve-EL+ managed to run all tests until termination or cancellation due to the time limit, whereas the blackbox variant using JCEL reproducibly failed for several test cases with $xT = 1000$ with 1 GB of heap space assigned. We conjecture that the large number of pre-solutions stored in this approach

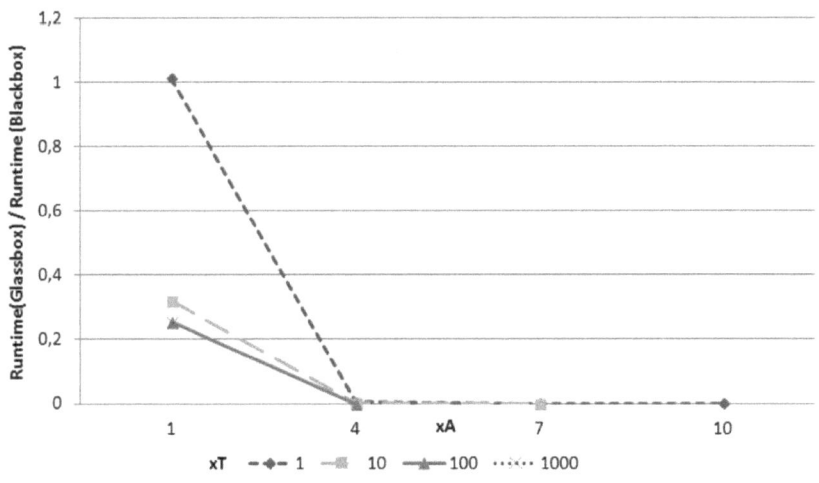

(a) xA values and relative runtime (split by xT)

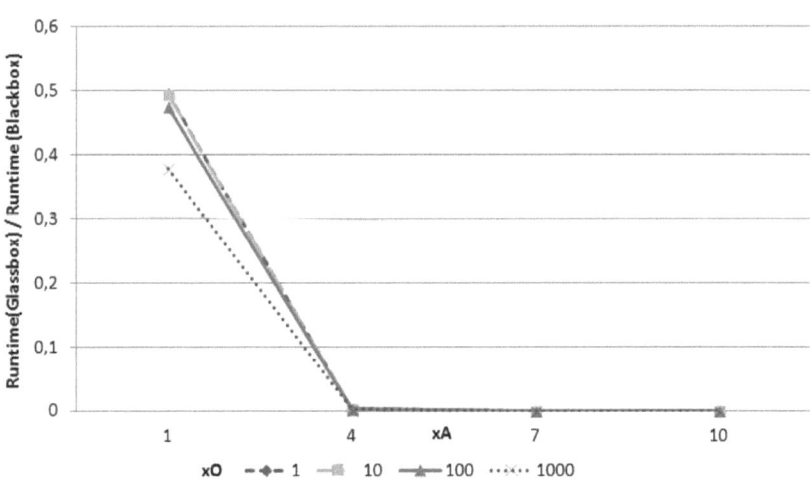

(b) xA values and relative runtime (split by xO)

Figure 4.1: Relative runtime of `RAPsolve-EL+` vs. `RAPsolve-Generic`

may be the reason for this behaviour and should be addressed in future work on a parallelised implementation.

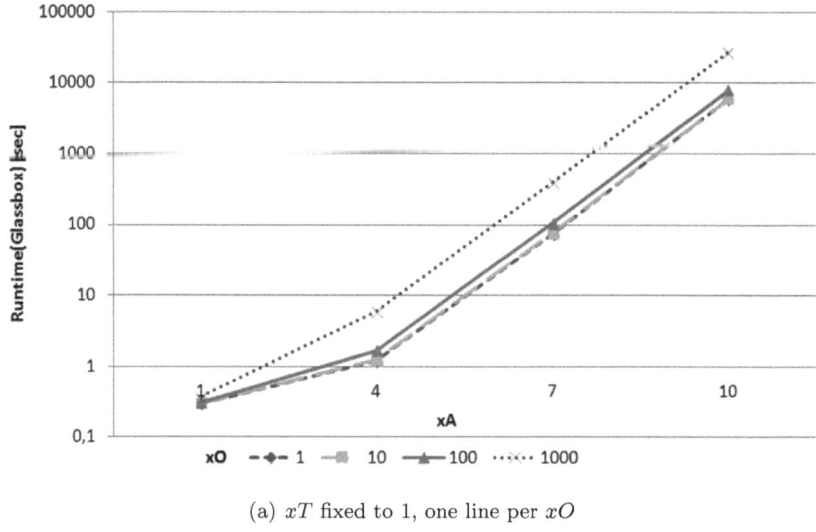

(a) xT fixed to 1, one line per xO

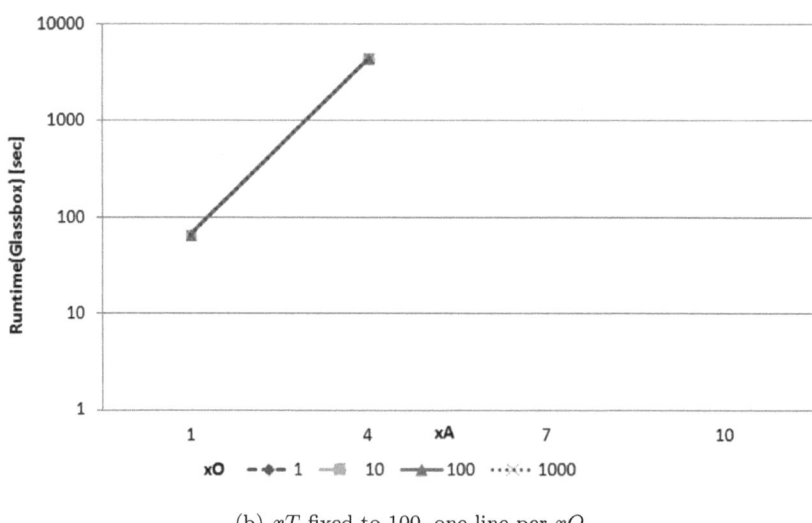

(b) xT fixed to 100, one line per xO

Figure 4.2: Runtime of `RAPsolve-EL+` vs. scaling factor xA

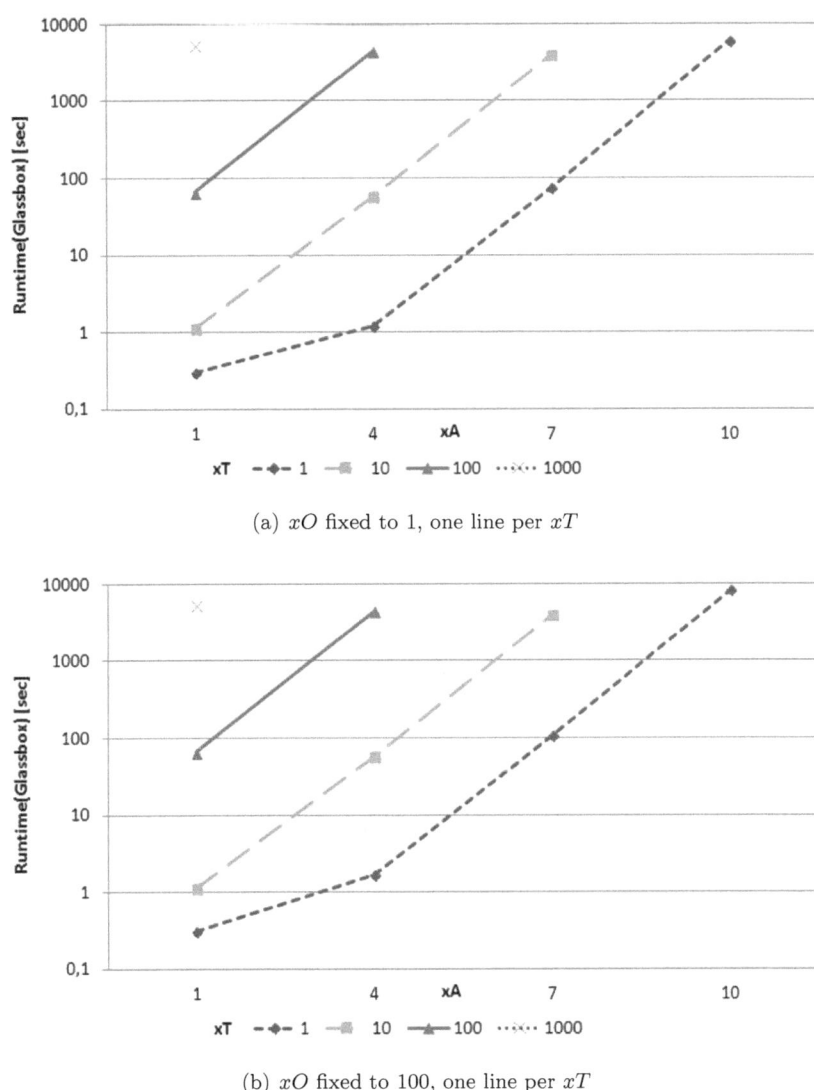

(a) xO fixed to 1, one line per xT

(b) xO fixed to 100, one line per xT

Figure 4.3: Runtime of `RAPsolve-EL+` vs. scaling factor xA

4.3 Application Scenarios and Expert Feedback

In the process of developing the relaxed abduction framework, we have been
guided by several real-world application scenarios. In order to validate the
theoretical results in an application context, and to gain valuable feedback
from domain experts on the practicability of the proposed solution, we have
conducted two small-scale experiments on selected scenarios jointly with three
domain experts. While this test group does not allow for any statistically
grounded conclusions, their feedback nevertheless gave a good overview on
potential strengths and weaknesses of relaxed abduction in related use cases.
Note that the focus of these experiments was not to assess usability of RAbIT,
or the comprehensibility of the reasoning mechanisms to potential users.
Rather, we wanted to understand if the typical problem sizes can be handled
by the current solution, and whether the results determined by relaxed
abduction do indeed represent added value for the subject matter expert.
The first application scenario addresses turbine diagnostics (characterized
by a typically small number of complex diagnoses, rather large domain
models, and a multitude of observations), whereas in the second scenario
we considered the task of rule-base debugging (with a typically much larger
number of rather simple hypotheses).

4.3.1 Scenario: Turbine Diagnostics

The turbine diagnostics scenario is structurally very similar to the running
example we used throughout this thesis to illustrate our ideas: Using the
terminology introduced in Section 3.1.2, the knowledge base $\mathcal{M}_{sys}$ is reused
without any modifications, whereas $\mathcal{M}_{cust}$ extends $\mathcal{M}_{sys}$ with the domain-
specific information on turbine structure (expressed in about 150 axioms).
As we employ (pure) causal modelling, $\mathcal{M}_{\overline{caus}}$ is empty, and its counterpart
$\mathcal{M}_{caus}$ comprises a total of 11 diagnostic axioms that relate the 11 key
diagnoses considered in the scenario to symptoms which can be observed in
the turbine's behaviour. Obviously, this scenario captures only a fraction
of what subject matter experts know about turbines. The main reason for
this restriction was the availability (or, rather, lack) of sufficiently formalized information available at the time of conducting the study: Structural
information on the turbines was mainly available as text documents and in
XML format, the main source for diagnostic knowledge were the experts
themselves. Formalizing this knowledge as ontologies was done manually for
the scope of our study; this task which will be continued in the context of
other activities, partially supported by scripts for converting the structural

information, and the use of abducible patterns (for more details see Hubauer et al., 2010) for generating $\mathcal{M}_{caus}$. Expressing observational information in $\mathcal{M}_{sym}$ is a straightforward task as long as symptom information explicitly contains the location (System) and was done using a simple script that creates appropriate axioms from the input.

Evaluation was conducted on a set of historical test cases comprising a set of component-symptom pairs and the diagnosis assigned by an expert at the time of occurrence. Each case was translated into an inclusion-based **sRAP** and passed to RAbIT for evaluation. The solutions were then presented to the experts ordered by decreasing coverage of the observation set.[4] As to be expected from the scalability results presented previously, performance was judged to be acceptable for offline processing. While scalability to larger sets of candidate diagnoses was judged crucial, one expert emphasized that in typical diagnostics scenarios, hypotheses often form a tree-like specialization structure which suggests an iterative approach where, initially, only the most general diagnoses (level 1) are taken as $\mathcal{A}$, and the solutions computed. In the next iteration, one could select the generic diagnosis that occurs most frequently in the solution set, and replace it by all its direct children in the diagnosis tree. Then, the solution set is recomputed or updated using the approach presented in Section 3.4.2. Analysing such an approach in detail will be part of future work. Considering the quality of the results generated, experts emphasized the significant added value of being able to provide simple explanations for most of the symptoms observed. In all cases but one, the expert-assigned "true" diagnosis was part of at least one of the computed solutions – in a large fraction of the cases, the diagnosis did even have the highest count of occurrences in the solution set among all candidates. Interestingly, it turned out that experts were especially interested in the observations not explained by one (or even all) solutions. Whereas observations not explained by an expert-selected preferred solution should be used as a starting point for additional analysis of the system, the occurrence of "unexplainable" observations seems to be a clear indicator for the need to adapt and improve the underlying domain formalization. When asked about additional features they would like to use in their daily work, the experts identified consideration of temporal dependencies of symptom occurrences as a key requirement for both online and offline application.

---

[4]Since the solution set was displayed in textual representation, experts were supported in reading them by the author.

4.3.2 Scenario: Debugging of Rule-Bases

The second evaluation scenario developed as a direct follow up to the discussions on turbine diagnostics: In one department, experts have compiled a large collection of rules in a domain-specific language, which they use to automate diagnostics for (a small portion of) turbine components. The rules formulate an anti-causal domain model for diagnostics (i. e. $S_1 \wedge \ldots \wedge S_n \rightarrow D_i$), and are evaluated by a simple, forward-chaining rule engine. While this straightforward solution does indeed support experts in many situations, it suffers from the incapability of handling observational incompleteness. In order to better understand if RAbIT can be of help here, we mapped the existing system into the relaxed abduction framework as follows: The structural component $\mathcal{M}_{sys}$ remains as is, and $\mathcal{M}_{cust}$ could also be taken over from the first scenario, despite the fact that due to the structure of the rule base, Symptoms are not assigned to components of the turbine, but to the System as a whole. This time $\mathcal{M}_{caus} = \emptyset$, and $\mathcal{M}_{\overline{caus}}$ contains one axiom of the form $\exists\,shows\,.\,S_1 \sqcap \ldots \sqcap \exists\,shows\,.\,S_n \sqsubseteq System \sqcap \exists\,influencedBy\,.$ $(\exists\,operatesIn\,.\,\langle D_i \rangle)$ for each rule in the knowledge base. Then, the hypotheses set $\mathcal{A}$ is comprised of (all) axioms stating the fact that a certain Symptom did occur, whereas axioms in $\mathcal{O}$ represent diagnoses that the engineer would expect to see produced by the rule engine in the situation at hand. Due to the "exchanged roles" of $\mathcal{A}$ and $\mathcal{O}$, it is not surprising to see that the according instantiation as a **sRAP** is typically characterized by a large set of hypotheses ($|\mathcal{A}| \geq 100$), whereas both the complexity of the hypotheses as well as the size of $\mathcal{O}$ are rather small. The size of $\mathcal{T}$ does not change significantly (although the axioms from $\mathcal{M}_{cust}$ could be removed as an optimization). Again, we used some rapidly prototyped scripts to create the sets $\mathcal{A}$ and $\mathcal{O}$ from a text file containing the test cases (this time comprised of a number of observed symptoms, and the diagnosis expected by the subject matter expert.

Surprisingly, a large fraction of the test cases could indeed be executed on our test system, though as an overnight task rather than interactively. There seems to be a multitude of reasons playing together to give this result: Firstly, the test system we used in this scenario was a server machine providing 512 GB of RAM and 8 CPUs with 4 cores each. Yet, RAbIT has not been implemented to make use of multi-core environments, which means that at most 2 of the cores have been used (as garbage collection is typically run in a separate thread). The reason for the runtime being orders of magnitude lower than expected therefore seems to lie in the problem structure itself: In contrast to `RAPsolve-Generic`, `RAPsolve-EL+` does not blindly test all possible assumptions, but only those whose preconditions are applicable in the current step (and only if no $\mathcal{T}$-axiom can be applied instead). Analysing a sample of the well-behaved cases together with an expert we found that either only a very small number of observations had been missing, or the number of relevant rules in the knowledge base was very small (sometimes as small as a single rule). So, while this scenario clearly revealed limits to the practical application of our approach, it also showed that there are "practically tractable" cases even for large hypothesis sets. A structured analysis of such cases should be conducted in the future. With regards to result quality, experts again made quite clear the potential benefit for their work. However, scalability plays an even more important role in this scenario, and further work will be needed.

5 Conclusion

The goal of this thesis as outlined in Section 1.2 has been to *develop a method for model-based information interpretation that addresses both observational incompleteness and incompleteness of the domain formalization at the same time, can be practically implemented, and easily applied in a wide range of industrial use cases.* We conclude this thesis by summing up the work done towards this goal, and the results achieved on the way. Moreover, we identify directions for future work that should be followed to improve upon the results achieved, wither scientifically or with regards to the practical applicability.

5.1 Discussion of Results

We defined the main scientific objective of this thesis to be the provisioning of *a formally founded framework for the interpretation of potentially incomplete information, guaranteeing both soundness and completeness of the result.* On the practical side, we took up the task to complement this framework *with an algorithm for effective and efficient generation of solutions* and, last but not least, with a methodology that connects the industry-relevant task of diagnostics of technical systems to the theoretical framework. We will now list the main contributions that we have made while working on this topic, both from a scientific and from an application-oriented perspective.

5.1.1 Scientific Contributions

Our academic key contribution is the definition and formalisation of relaxed abduction as a generic framework for interpreting information that conservatively extends standard logic based abduction.[1] We have shown that this formalisation is flexible enough to handle incompleteness of observations and domain representation in an integrated way without demanding commensurability of the two. Also, we conducted a detailed analysis of the formal

---

[1] In fact, a number of alternative abduction frameworks and even pinpointing were identified as special cases of relaxed abduction.

properties of relaxed abduction, and identified the important class of strict relaxed abduction problems for which solutions correspond to Pareto-optimal elements of the solution space, and derived a $(O(2^{|A|}))$ upper bound on the number of solutions.

Next, we analysed how concrete choices of the preference relations $\preceq_A$ and $\preceq_O$ affect the properties of a relaxed abduction problem – first and foremost the possible size of the solution set. Subset-based and entailment-based orders were identified as especially promising candidates for which the expected solution count is reduced to $O(1.5^{|A|})$. In addition to providing two algorithms for solving relaxed abduction problems (one optimized for $\mathcal{EL}^+$, the other generically applicable to any DL), we integrated recent results from other researchers to outline extensions of the basic solution towards more expressive underlying logics, and the support for incremental reasoning. Although some results from literature, such as ATMSs, seem to provide similar features to the proposed solution, the novelty of our approach was successfully validated in a comparative analysis against the current state of the art, based on the requirements defined at the beginning of our work.

5.1.2 Practical Applications

The major practical result of this work is the reasoning engine RAbIT which implements both the generic blackbox algorithm as well as the $\mathcal{EL}^+$-specific glassbox variant to solve relaxed abduction problems. Based on this proto-typical implementation, we conducted a systematic performance evaluation which showed that despite the reduction in solution space induced by the problem structure, relaxed abduction is and remains a challenging and com-putationally hard problem (in fact, theory shows the problem is indeed EXPTIME-hard). However, especially for situations where the number of hypotheses is small (for instance, 10 to 15), relaxed abduction problems can efficiently be solved by the current implementation. Workshops conduc-ted together with subject matter experts showed that, given appropriate modelling of the problem, this performance is indeed sufficient to support those experts in several of their daily tasks, given an apropriate modelling of the domain. While the potential for performance improvements neverthe-less became clear, the experts gave very positive feedback on the general idea of relaxed abduction, i. e. allowing explanations omit certain observed facts if this makes them simpler. Overall, we consider our initial evaluation successful.

Another applied result of the thesis are the mappings into the frame-work of relaxed abduction we defined for a number of different information interpretation problems, most prominently for diagnostic reasoning using causal domain models. Providing such an integrated methodology that combines modelling and reasoning in an extensible manner is a significant step to strengthen the link between our novel formalism and real industrial applications.

Summarising our contributions, this thesis makes a significant step in closing the gap between theoretical results in the field of knowledge representation and reasoning, and their practical applicability in the field of industrial information interpretation.

5.2 Directions for Future Work

A number of interesting questions and ideas could not be addressed in the scope of this thesis. However, following up on them should add significant value to the relaxed abduction framework in general, and its implementation in RAbIT.

Firstly, the development of modern description logics reasoners strikingly demonstrates that theoretical ExpTime-hardness of a problem does not mean that scalable implementations are not feasible. This should encourage additional work to further improve the scalability of the current implementation: Parallelisation is a natural choice especially in the case of RAPsolve-Generic since the runs are independent, but should also be considered for RAPsolve-EL+, as suggested by recent work on parallel classification by Kazakov et al. (2011).

Moreover, complex problems tend to have special cases whose complexity is much lower – PTime-reasoning for $\mathcal{EL}^+$ being a prominent example. In Section 4.3 we already pointed out that there seem to be specific structures in a **RAP** that reduce computation time: Loosely-connected knowledge bases, potentially exhibiting an island-structure as described by Wandelt & Möller (2010), could result in even lower complexity bounds. Similarly, diagnoses are often formulated from general to specific, which may give rise to a specialization structure between hypotheses. Further research should be conducted to automatically detect and benefit from such structures. We also saw that the focus on (dually) irredundant preorders (and, therefore, strict **RAPs**), allowed us to define stronger results on the maximum label size. An investigation of other classes of preorders might identify ways to strengthen these results even more.

Third and last, expert feedback included relatively concrete feature requests, namely support for expressing and reasoning over temporal dependencies, and support for the handling of unexplained observations (i. e. observations not explained by any preferred solution). Whereas the latter is more of a software engineering task, the former tackles the field of temporal description logics as outlined by Artale & Franconi (2000). Their combination with (relaxed) abduction may pose challenging new questions.

Bibliography

Allemang, D., Tanner, M. C., Bylander, T., & Josephson, J. R. (1987). Computational complexity of hypothesis assembly. In J. P. McDermott (Ed.), *Proceedings of the 10th International Joint Conference on Artificial Intelligence*, volume 2 (pp. 1112–1119). Milan, Italy: Morgan Kaufmann. (2 citations on pages 103 and 108)

Appelt, D. E. & Pollack, M. E. (1992). Weighted abduction for plan ascription. In *User Modeling and User-Adapted Interaction*, volume 2 (pp. 1–25).: Springer. (cited on page 12)

Artale, A. & Franconi, E. (2000). A survey of temporal extensions of description logics. *Annals of Mathematics and Artificial Intelligence (AMAI)*, 30(1-4), 171–210. (cited on page 124)

Baader, F. (2003). Terminological cycles in a description logic with existential restrictions. In G. Gottlob & T. Walsh (Eds.), *Proceedings of the 18th International Joint Conference on Artificial Intelligence (IJCAI 2003)* (pp. 325–330). Acapulco, Mexico: Morgan Kaufmann. (cited on page 56)

Baader, F., Brandt, S., & Lutz, C. (2005a). Pushing the EL envelope. In L. P. Kaelbling & A. Saffiotti (Eds.), *Proceedings of the 19th International Joint Conference on Artificial Intelligence (IJCAI 2005)* (pp. 364–369). Edinburgh, Scotland, UK: Professional Book Center. (5 citations on pages 11, 56, 66, and 83)

Baader, F., Brandt, S., & Lutz, C. (2005b). *Pushing the EL Envelope.* LTCS-Report LTCS-05-01, Institute for Theoretical Computer Science, TU Dresden. (5 citations on pages 56, 57, 58, 60, and 83)

Baader, F., Peñaloza, R., & Suntisrivaraporn, B. (2007). Pinpointing in the description logic EL+. In J. Hertzberg, M. Beetz, & R. Englert (Eds.), *Proceedings of the 30th Annual German Conference on AI (KI 2007)*, volume 4667 of *Lecture Notes in Computer Science* (pp. 52–67). Osnabrück, Germany: Springer. (5 citations on pages 11, 50, 58, 106, and 108)

Daget, J.-F., Leclère, M., & Salvat, E. (2011). On rules with existential variables: Walking the decidability line. *Artificial Intelligence*, 175(9-10), 1620–1654. (cited on page 104)

Baral, C. (2000). Abductive reasoning through filtering. *Artificial Intelligence*, 120(1), 1–28. (2 citations on pages 103 and 108)

Bentley, J. L., Kung, H. T., Schkolnick, M., & Thompson, C. D. (1978). On the average number of maxima in a set of vectors and applications. *Journal of the ACM*, 25(4). (cited on page 79)

Bienvenu, M. (2008). Complexity of abduction in the EL family of lightweight description logics. In G. Brewka & J. Lang (Eds.), *Proceedings of the 11th International Conference on Principles of Knowledge Representation and Reasoning (KR 2008)* (pp. 220–230). Sydney, Australia: AAAI Press. (4 citations on pages 15, 38, 103, and 108)

Bochman, A. (2007). A causal theory of abduction. *Journal of Logic and Computation*, 17, 851–869. (cited on page 19)

Booth, R. & Meyer, T. (2011). How to revise a total preorder. *Journal of Philosophical Logic*, 40(2), 193–238. (cited on page 23)

Brazier, F., Treur, J., & Wijngaards, N. (1996). The acquisition of a shared task model. In N. Shadbolt, K. O'Hara, & G. Schreiber (Eds.), *Proceedings of the 9th European Knowledge Acquisition Workshop on Advances in Knowledge Akquisition (EKAW 1996)*, volume 1076 of *Lecture Notes in Computer Science* (pp. 278–289).: Springer. (cited on page 29)

Buss, S. R. (1998). *An Introduction to Proof Theory*, chapter 1, (pp. 1–78). Elsevier. (2 citations on pages 47 and 64)

Castano, S., Espinosa Peraldi, I. S., Ferrara, A., Karkaletsis, V., Kaya, A., Möller, R., Montanelli, S., Petasis, G., & Wessel, M. (2009). Multimedia interpretation for dynamic ontology evolution. *Journal of Logic and Computation*, 19(5), 859–897. (cited on page 17)

Cherkassky, B. V., Goldberg, A. V., & Radzik, T. (1996). Shortest paths algorithms: Theory and experimental evaluation. *Mathematical Programming*, 73(2), 129–174. (cited on page 66)

Clark, K. L. (1977). Negation as failure. In H. Gallaire & J. Minker (Eds.), *Logic and Data Bases*, volume 0 of *Advances in Data Base Theory* (pp. 293–322).: Plemum Press. (cited on page 32)

Colucci, S., Di Noia, T., Di Sciascio, E., Donini, F. M., & Mongiello, M. (2003). Concept abduction and contraction in description logics. In D. Calvanese, G. De Giacomo, & E. Franconi (Eds.), *Proceedings of the 16th International Workshop on Description Logics (DL 2003)*, volume 81 of *CEUR Workshop Proceedings* Rome, Italy: CEUR-WS.org. (cited on page 16)

Console, L., Dupré, D. T., & Torasso, P. (1991). On the relationship between abduction and deduction. *Journal of Logic and Computation*, 1(5), 661–690. (cited on page 51)

De Kleer, J. (1986). An assumption-based TMS. *Artificial Intelligence*, 28(2), 127–162. (2 citations on pages 58 and 107)

De Kleer, J. & Kurien, J. (2003). Fundamentals of model-based diagnosis. In *Proceedings of the 5th IFAC Symposium on Fault Detection, Supervision and Safety of Technical Processes (Safeprocess 2003)*. (2 citations on pages 54 and 104)

Denecker, M. & Kakas, A. C. (2002). Abduction in logic programming. In A. C. Kakas & F. Sadri (Eds.), *Computational Logic. Logic Programming and Beyond*, volume 2407 of *Lecture Notes in Computer Science* (pp. 402–436). Springer. (cited on page 104)

Dressler, O. & Puppe, F. (1999). Knowledge-based diagnosis - survey and future directions. In F. Puppe (Ed.), *Proceedings of the 5th Biannual German Conference on Knowledge-Based Systems (XPS-99)*, volume 1570 of *Lecture Notes in Computer Science* (pp. 24–46). Würzburg, Germany: Springer. (2 citations on pages 4 and 106)

Eiter, T. & Gottlob, G. (1995). The complexity of logic-based abduction. *Journal of the ACM*, 42(1), 3–42. (5 citations on pages 14, 35, 38, 45, and 103)

Fellner, W. (1968). *An Introduction to Probability Theory and its Applications*, volume 1. Wiley, 3rd edition. (cited on page 42)

Forbus, K. D. & De Kleer, J. (1993). *Building Problem Solvers*. Artificial Intelligence Series. MIT Press. (2 citations on pages 107 and 108)

Freeman, J. W. (1995). *Improvements to Propositional Satisfiability Search Algorithms*. Phd thesis, University of Pennsylvania. (cited on page 58)

Gentzen, G. (1934). Untersuchungen über das logische Schließen I. *Mathematische Zeitschrift*, 39(2), 176–210. In German. (cited on page 55)

Gentzen, G. (1935). Untersuchungen über das logische Schließen II. *Mathematische Zeitschrift*, 39(3), 405–431. In German. (cited on page 55)

Grau, B. C., Halaschek-Wiener, C., Kazakov, Y., & Suntisrivaraporn, B. (2010). Incremental classification of description locigs ontologies. *Journal of Automated Reasoning*, 44(4), 337–369. (4 citations on pages 93, 95, and 99)

Gries, O., Möller, R., Nafissi, A., Rosenfeld, M., Sokolski, K., & Wessel, M. (2010). A probabilistic abduction engine for media interpretation based on ontologies. In T. Lukasiewicz, R. Peñaloza, & A.-Y. Turhan (Eds.), *Proceedings of the 1st International Workshop on Uncertainty in Description Logics (UniDL 2010)*, volume 613 of *CEUR Workshop Proceedings* Edinburgh, UK: CEUR-WS.org. (3 citations on pages 104, 105, and 108)

Grimm, S., Watzke, M., Hubauer, T. M., & Cescolini, F. (2012). Embedded EL+ reasoning on programmable logic controllers. In P. Cudré-Mauroux, J. Heflin, E. Sirin, T. Tudorache, J. Euzenat, M. Hauswirth, J. X. Parreira, J. Hendler, G. Schreiber, A. Bernstein, & E. Blomqvist (Eds.), *Proceedings of the 11th International Semantic Web Conference (ISWC 2012) - Part II*, volume 7650 of *Lecture Notes in Computer Science* (pp. 66–81). Boston, USA: Springer. (cited on page 6)

Grimm, S. & Wissmann, J. (2011). Elimination of redundancy in ontologies. In M. Grobelnik & E. Simperl (Eds.), *Proceedings of the 8th European Semantic Web Conference (ESWC 2011)*, volume 6643 of *Lecture Notes in Computer Science* (pp. 260–274).: Springer. To appear. (cited on page 10)

Guerriero, F. & Musmanno, R. (2001). Label correcting methods to solve multicriteria shortest path problems. *Journal of Optimization Theory and Applications*, 111(3), 589–613. (3 citations on pages 66, 73, and 74)

Halaschek-Wiener, C. & Katz, Y. (2006). Belief base revision for expressive description logics. In B. C. Grau, P. Hitzler, C. Shankey, & E. Wallace (Eds.), *Proceedings of the OWLED 2006 Workshop on OWL: Experiences and Directions*, volume 216 of *CEUR Workshop Proceedings* Athens, Georgia, USA: CEUR-WS.org. (cited on page 94)

Hartshorne, C. & Weiss, P., Eds. (1931). *Collected Papers of Charles Sanders Peirce*. Harvard University Press, 1st edition. (cited on page 12)

Hobbs, J. R., Stickel, M. E., Appelt, D. E., & Martin, P. A. (1993). Interpretation as abduction. *Artificial Intelligence*, 63(1-2), 69–142. (6 citations on pages 12, 33, 34, 103, and 108)

Horridge, M. & Bechhofer, S. (2009). The OWL API: A java API for working with OWL 2 ontologies. In R. Hoekstra & P. F. Patel-Schneider (Eds.), *Proceedings of the 6th International Workshop on OWL: Experiences and Directions (OWLED 2009)*, volume 529 of *CEUR Workshop Proceedings* Chantilly, Virginia, USA: CEUR-WS.org. (cited on page 109)

Horrocks, I. R., Kutz, O., & Sattler, U. (2006). The even more irresistible SROIQ. In P. Doherty, J. Mylopoulos, & C. A. Welty (Eds.), *Proceedings of the 10th International Conference on Principles of Knowledge Representation and Reasoning (KR 2006)* (pp. 57–67). Lake District, United Kingdom: AAAI Press. (cited on page 11)

Hubauer, T. M., Grimm, S., Lamparter, S., & Roshchin, M. (2012). A diagnostics framework based on abductive description logic reasoning. In *Proceedings of the 2012 IEEE International Conference on Industrial Technology*: IEEE Computer Society. Accepted for publication. (2 citations on pages 6 and 51)

Hubauer, T. M., Lamparter, S., & Pirker, M. (2010). Automata-based abduction for tractable diagnosis. In V. Haarslev, D. Toman, & G. E. Weddell (Eds.), *Proceedings of the DL Home 23rd International Workshop on Description Logics (DL 2010)*, volume 573 of *CEUR Workshop Proceedings* (pp. 360–371). Waterloo, CA: CEUR-WS.org. (2 citations on pages 7 and 117)

Hubauer, T. M., Lamparter, S., & Pirker, M. (2011a). Relaxed abduction: Robust information interpretation for incomplete models. In R. Rosati, S. Rudolph, & M. Zakharyaschev (Eds.), *Proceedings of the DL Home 24th International Workshop on Description Logics (DL 2011)*, volume 745 of *CEUR Workshop Proceedings* (pp. 180–190). Barcelona, Spain: CEUR-WS.org. (cited on page 7)

Hubauer, T. M., Legat, C., & Seitz, C. (2011b). Empowering adaptive manufacturing with interactive diagnostics: A multi-agent approach. In *Proceedings of the 9th International Conference on Practical Applications of Agents and Multi-Agent Systems*, volume 88 of *Advances in Intelligent and Soft-Computing* (pp. 47–56).: Springer. (4 citations on pages 6, 17, and 27)

ISO (2003). *ISO 13379:2003: Condition Monitoring and Diagnostics of Machines - General Guidelines on Data Interpretation and Diagnostics Techniques.* Geneva, Switzerland: International Organization for Standardization. (cited on page 28)

Kate, R. J. & Mooney, R. J. (2009). Probabilistic abduction using markov logic networks. In *Proceedings of the IJCAI-09 Workshop on Plan, Activity and Intent Recognition (PAIR 2009)* Pasadena, California, USA. (cited on page 106)

Kaya, A. (2011). *A Logic-Based Approach to Multimedia Interpretation.* Phd thesis, Hamburg University of Technology. (cited on page 19)

Kazakov, Y. (2008). SRIQ and SROIQ are harder than SHOIQ. In G. Brewka & J. Lang (Eds.), *Proceedings of the 11th International Conference on Principles of Knowledge Representation and Reasoning (KR 2008)* (pp. 274–284). Sydney, Australia: AAAI Press. (cited on page 11)

Kazakov, Y. (2009). Consequence-driven reasoning for horn SHIQ ontologies. In B. C. Grau, I. R. Horrocks, B. Motik, & U. Sattler (Eds.), *Proceedings of the DL Home 22nd International Workshop on Description Logics (DL 2009)*, volume 477 of *CEUR Workshop Proceedings* Oxford, UK: CEUR-WS.org. (2 citations on pages 87 and 92)

Kazakov, Y., Krötzsch, M., & Simančík, F. (2011). Concurrent classification of EL ontologies. In L. Aroyo, C. A. Welty, H. Alani, J. Taylor, A. Bernstein, L. Kagal, N. Fridman Noy, & E. Blomqvist (Eds.), *Proceedings of the 10th International Semantic Web Conference (ISWC 2011)*, volume 7031 of *Lecture Notes in Computer Science* (pp. 305–320). Bonn, Germany: Springer. (cited on page 123)

Klarmann, S., Endriss, U., & Schlobach, S. (2011). Abox abduction in the description logic ALC. *Journal of Automated Reasoning*, 46(1), 43–80. (2 citations on pages 105 and 108)

Kowalski, R. A. (2011). *Computational Logic and Human Thinking: How to be Artificially Intelligent.* Cambridge University Press, 1st edition. (4 citations on pages 31, 32, 105, and 108)

Legat, C., Hubauer, T. M., & Seitz, C. (2011). Integrated diagnosis for adaptive service-oriented manufacturing control with autonomous products. In L. Benyoucef, D. Trentesaux, A. Artiba, & N. Rezg (Eds.), *Proceedings of the 2011 International Conference on Industrial Engineering and Systems Management (IESM 2011)* (pp. 1363–1372). Metz, France: École nationale d'ingénieurs de Metz (ENIM) International Institute for Innovation, Industrial Engineering and Entrepreneurship (I^{4}e^2). (2 citations on pages 6 and 27)

Mendez, J. (2012). jCel: A modular rule-based reasoner. In *Proceedings of the 1st International Workshop on OWL Reasoner Evaluation (ORE 2012)*. (cited on page 110)

Mendez, J., Ecke, A., & Turhan, A.-Y. (2011). Implementing completion-based inferences for the EL family. In R. Rosati, S. Rudolph, & M. Zakharyaschev (Eds.), *Proceedings of the DL Home 24th International Workshop on Description Logics (DL 2011)*, volume 745 of *CEUR Workshop Proceedings* (pp. 334–344). Barcelona, Spain: CEUR-WS.org. (cited on page 106)

Möller, R. & Neumann, B. (2008). *Ontology-Based Reasoning Techniques for Multimedia Interpretation and Retrieval.* Springer. (cited on page 12)

Nafissi, A. (2013). *Applying Markov Logics for Controlling Abox Abduction.* PhD thesis, Hamburg University of Technology. (4 citations on pages 20, 104, 105, and 108)

Nardi, D., Brachman, R. J., Baader, F., Nutt, W., Donini, F. M., Sattler, U., Calvanese, D., Molitor, R., de Giacomo, G., Küsters, R., Wolter, F., McGuinness, D. L., Patel-Schneider, P. F., Möller, R., Haarslev, V., Horrocks, I. R., Borgida, A., Welty, C. A., Rector, A. L., Franconi, E., Lenzerini, M., & Rosati, R. (2003). *The Description Logic Handbook: Theory, Implementation, and Applications.* Cambridge University Press. (2 citations on pages 9 and 11)

Nielsen, L. R. (2001). *A Bicriterion and Parametric Analysis of the Shortest Hyperpath Problem.* Progress report, University of Aarhus - Department of Operations Research. (cited on page 24)

Norvig, P. & Wilensky, R. (1990). A critical evaluation of commensurable abduction models for semantic interpretation. In *Proceedings of the 13th Conference on Computational Linguistics (COLING 1990)* (pp. 225–230).

Helsinki, Finland: Association for Computational Linguistics. (cited on page 34)

Paul, G. (1993). Approaches to abductive reasoning: An overview. *Artificial Intelligence Review*, 7(2), 109–152. (3 citations on pages 13, 107, and 108)

Peraldi, I. S. E. (2011). *Content Management and Knowledge Management:: Two Faces of Ontology-based Deep-Level Interpretation of Texts*. Phd thesis, Hamburg University of Technology. (cited on page 19)

Peraldi, I. S. E., Kaya, A., Melzer, S., Möller, R., & Wessel, M. (2007). Towards a media interpretation framework for the semantic web. In *Proceedings of the 2007 IEEE / WIC / ACM International Conference on Web Intelligence (WI 2007)* Silicon Valley, California, USA: IEEE Computer Society. (3 citations on pages 12, 104, and 108)

Peraldi, I. S. E., Kaya, A., & Möller, R. (2009). Formalizing mulimedia interpretation based on abduction over description logic aboxes. In B. C. Grau, I. R. Horrocks, B. Motik, & U. Sattler (Eds.), *Proceedings of the DL Home 22nd International Workshop on Description Logics (DL 2009)*, volume 477 of *CEUR Workshop Proceedings* Oxford, UK: CEUR-WS.org. (4 citations on pages 104, 105, and 108)

Poole, D. (1988). Representing knowledge for logic-based diagnosis. In I. for New Generation Computer Technology (ICOT) (Ed.), *Proceedings of the International Conference on Fifth Generation Computer Systems (FGCS 1988)* Tokyo, Japan: Springer. (5 citations on pages 29, 105, and 108)

Pople, H. E. (1973). On the mechanization of abductive logic. In N. J. Nilsson (Ed.), *Proceedings of the 3rd International Joint Conference on Artificial Intelligence (IJCAI 1973)* (pp. 147–152). Stanford, California, USA: Morgan Kaufmann. (cited on page 12)

Reggia, J. A., Nau, D. S., & Wang, P. Y. (1983). Diagnostic expert systems based on a set covering model. *International Journal of Man-Machine Studies*, 19(5), 437–460. (2 citations on pages 103 and 108)

Reiter, R. & De Kleer, J. (1988). Foundations of assumption-based truth maintenance systems: Preliminary report. In *Proceedings of the 7th National Conference on Artificial Intelligence (AAAI 1987)* (pp. 183–189). St. Paul, Minnesota, USA: AAAI Press / The MIT Press. (cited on page 107)

Scott, D. (1974). Completeness and axiomatizability in many-valued logic. In *Proceedings of the Tarski Symposium: Proceedings of Symposia in Pure Mathematics*, volume 25 (pp. 411–435).: American Mathematical Society. (cited on page 19)

Selman, B. & Levesque, H. J. (1990). Abductive and default reasoning: A computational core. In *Proceedings of the 8th National Conference on Artificial Intelligence (AAAI 1990)* (pp. 343–348). Boston, Massachusetts, USA: AAAI Press / The MIT Press. (cited on page 39)

Shanahan, M. (2005). Perception as abduction: Turning sensor data into meaningful representation. *Cognitive Science*, 29(1), 103–134. (cited on page 12)

Sirin, E., Parsia, B., Grau, B. C., Kalyanpur, A., & Katz, Y. (2007). Pellet: A practical OWL-DL reasoner. *Web Semantics: Science, Services and Agents on the World Wide Web*, 5(2), 51–53. Software Engineering and the Semantic Web. (cited on page 110)

Skriver, A. J. V. (2000). A classification of bicriterion shortest path (BSP) algorithms. *Asia-Pacific Journal of Operational Research*, 17, 199–212. (3 citations on pages 66, 67, and 73)

Sperner, E. (1928). Ein Satz über Untermengen einer endlichen Menge. *Mathematische Zeitschrift*, 27, 544–548. In German. (cited on page 42)

Struss, P. (1997). Model-based and qualitative reasoning: An introduction. *Annals of Mathematics and Artificial Intelligence (AMAI)*, 19(3-4), 355–381. (2 citations on pages 55 and 104)

Thagard, P. R. (1978). The best explanation: Criteria for theory choice. *Journal of Philosophy*, 75(2), 76–92. (cited on page 33)

W3C OWL Working Group (2009a). *OWL 2 Web Ontology Language: Primer*. W3C Recommendation. (cited on page 11)

W3C OWL Working Group (2009b). *OWL 2 Web Ontology Language: Profiles*. W3C Recommendation. (2 citations on pages 11 and 87)

Wandelt, S. & Möller, R. (2010). Distributed island-based query answering for expressive ontologies. In V. Haarslev, D. Toman, & G. E. Weddell (Eds.), *Proceedings of the DL Home 23rd International Workshop on Description Logics (DL 2010)*, volume 573 of *CEUR Workshop Proceedings* (pp. 185–196). Waterloo, CA: CEUR-WS.org. (cited on page 123)

Zadeh, L. A. (1965). Fuzzy sets. *Information and Control*, 8(3), 338–353. (cited on page 49)